GENETICALLY EDITED CHILDREN AND THE RISKS TO EQUALITY

Edited by
Calum MacKellar, Roberto Andorno,
and Matthew James

BRISTOL
UNIVERSITY
PRESS

First published in Great Britain in 2026 by

Bristol University Press
University of Bristol
1–9 Old Park Hill
Bristol
BS2 8BB
UK
t: +44 (0)117 374 6645
e: bup-info@bristol.ac.uk

Details of international sales and distribution partners are available at bristoluniversitypress.co.uk

DOI: 10.51952/9781529249552

British Library Cataloguing in Publication Data
A catalogue record for this book is available from the British Library

ISBN 978-1-5292-4953-8 hardcover
ISBN 978-1-5292-4954-5 ePub
ISBN 978-1-5292-4955-2 ePdf

Cover design: Newgen
Front cover image: Stocksy/The Laundry Room
Bristol University Press uses environmentally responsible print partners.
Printed and bound in Great Britain by CPI Group (UK) Ltd, Croydon, CR0 4YY

FSC
www.fsc.org
MIX
Paper | Supporting responsible forestry
FSC® C013604

Bristol University Press' authorised representative in the European Union is:
Easy Access System Europe, Mustamäe tee 50, 10621 Tallinn, Estonia,
Email: gpsr.requests@easproject.com

Contents

Notes on Contributors

Roberto Andorno is Associate Professor in the Faculty of Law and a Research Associate in the Institute of Biomedical Ethics and History of Medicine at the University of Zurich, Switzerland.

Vicente Bellver is Professor in the Department of Legal and Political Philosophy at the University of Valencia, Spain.

Rosemarie Garland-Thomson is Professor Emerita of English and Bioethics at Emory University in Atlanta, Georgia, USA.

Katie Hasson is Executive Director of the Center for Genetics and Society, California, USA. She was previously Assistant Professor of Sociology and Gender Studies at the University of Southern California, USA.

J. Benjamin Hurlbut is Associate Professor in the School of Life Sciences at Arizona State University, USA and Codirector of the Global Observatory on Genome Editing.

Laura E. Jacobs is Doctoral Candidate at the Department of Legal Theory, Faculty of Law at the Free University of Amsterdam, The Netherlands. Her research focuses on the legal and philosophical dimensions of social decision-making of human germline gene editing.

Matthew James is Associate Professor of Bioethics in the School of Theology and Director of the Centre for Bioethics and Emerging Technologies at St Mary's University, England.

David A. Jones is Professor of Bioethics at St Mary's University, Twickenham, and Research Fellow at Blackfriars Hall at the University of Oxford, England.

Calum MacKellar is Director of Research of the Scottish Council on Human Bioethics in Edinburgh, Scotland, and Visiting Lecturer in Bioethics at St Mary's University, England.

Pia Matthews is Senior Lecturer in Bioethics in the School of Theology at St Mary's University, England.

Richard Playford is Senior Lecturer in Philosophy, Ethics and Religion at Leeds Trinity University, England.

Sandy Sufian is Professor in the Department of Medical Education and in the Department of Disability Studies and Human Development at the University of Illinois Chicago, USA.

Trevor Stammers is now retired. He was previously a physician for over 30 years and Associate Professor in Bioethics. He is a Member of the Scottish Council on Human Bioethics, Edinburgh, Scotland.

Britta van Beers is Professor in the Department of Legal Theory, Faculty of Law at the Free University of Amsterdam, The Netherlands.

Acknowledgements

This book was prepared in collaboration with the following organizations:

1. The Scottish Council on Human Bioethics (SCHB) in Edinburgh, Scotland, which was formed in 1997 as an independent registered Scottish charity made up of physicians, lawyers, ethicists, and other professionals from disciplines associated with medical ethics. The principles to which the SCHB subscribes are set out in the 1948 United Nations Universal Declaration of Human Rights.
2. The Centre for Bioethics and Emerging Technologies, which was created in 2009 as part of St Mary's University in London, England, which focuses on the ethical and social implications of biomedicine and other new technologies. It coordinates postgraduate research and organizes conferences as well as regular series of public seminars on major issues in bioethics.
3. The Social Trends Institute in Barcelona, Spain, which is an independent, non-profit research centre that offers institutional and financial support to academics who seek to make sense of emerging social trends and their effects on human communities.

Introduction

Calum MacKellar, Roberto Andorno, and Matthew James

A new era in genome editing

In the past, modifying (editing) the DNA of human cells was technically challenging, time-consuming, and expensive.[1] In 2014, however, a revolution took place when biological scientists Jennifer Doudna from the USA and Emmanuelle Charpentier from France announced that they had discovered a completely new system to modify genes called CRISPR/Cas9 for which they received the 2020 Nobel Prize for Chemistry.[2] This was a very much simpler, faster, cheaper, and more efficient procedure than older methods. The ease of use, together with the high specificity of the CRISPR/Cas9 complex, has transformed the field of editing the genes of many different organisms, including human beings (National Academies of Sciences, Engineering, and Medicine, 2017).

This new genome (complete genetic heritage of a living being) editing system was adapted from a natural defence mechanism used by bacteria. When these are infected by viruses, they capture small segments of the viruses' DNA and integrate them into their own DNA in a specific pattern, creating what are known as CRISPR arrays. These enable the bacteria to 'remember' the viruses that previously attacked them. If the same virus attacks again, the bacteria produce RNA segments from the CRISPR arrays that can recognize and bind to specific regions of the viral DNA. The bacteria then utilize the Cas9 enzyme (a biological catalyst) to cut the viral DNA and disable the virus. This bacterial defence system is now used by scientists for the purpose of editing the DNA of any living organism, including in human research where specific gene sequences may be disabled or replaced for potential therapies (Li et al, 2020) and for better understanding embryonic development (Zhang et al, 2023). However, it has also led to significant ethical debate about the possible use of certain kinds of genome editing procedures in future human beings (Evans, 2020).

The crucial distinction between somatic and heritable genome editing

In this context, it is crucial to distinguish between the genetic genome editing of somatic (non-reproductive) cells and that of sperm and egg cells as well as early embryos (Sparrow, 2022; MacKellar, 2023).

Somatic gene editing

With somatic cell gene editing it may be possible to address a genetic disorder on a mature embryo, foetus, child, or adult by inactivating disordered genes which may be responsible for a disease. Gene editing could then be considered for new treatments of genetic disorders in individuals (Sas and Lawrenz, 2017), without intentionally affecting the genome of their descendants. This means that such a procedure could be considered in a similar manner to already existing somatic gene therapy procedures which have generally been accepted and welcomed by the whole of society. In addition, they would not raise any new significant ethical problems, apart from safety and efficacy (MacKellar, 2017). For example, a new somatic gene therapy using gene editing has already been developed for Sickle cell disease, which is a rare, debilitating, and life-threatening blood disorder (Dimitrievska et al, 2024).

Moreover, since somatic gene editing would be a form of therapy (or treatment) it is also considered to be a medical intervention, meaning that it restores an existing human patient's functioning to species-typical norms or that it gives abilities integral to the body which are considered to be normal. A therapy thus counteracts a known or an anticipated health deficit (British Medical Association, 2007, p 5).

Theoretically, somatic gene editing could also be used for enhancement (or augmentation) purposes which includes activities seeking to increase the human functioning of a *healthy* individual beyond the norm for that person and in the absence of any identified dysfunction (Harris, 2007, p 9). In other words, it can be defined as an activity designed 'to alter … the "normal" workings of the human body and psyche, to augment or improve their capacities and performances' (President's Council, 2003, p 13). Generally, the aim of enhancement is to improve upon the norm but not to surpass a pre-existing, human, natural state or capacity (MacKellar and Bechtel, 2014, pp 5–6).

Sometimes, however, distinguishing 'therapy' from 'enhancement' is difficult and depends on the definitions of other terms as well as cultural norms and values. For example, if a society willingly seeks to enhance its members, then what would be considered normal for this community would eventually be altered. Previously normal traits could even be considered as dysfunctional if they no longer attain the new 'norm'. In such an event, these new dysfunctions could begin to be considered for treatment (MacKellar and Bechtel, 2014, pp 5–6).

Gene editing of very early embryos

When gene editing is performed on a very early-stage human embryo, such as at the two-cell stage, several ethical concerns arise. A key issue is determining whether substantial genetic modifications result in the creation of an entirely new individual or whether the original embryo persists but undergoes modification (Ossorio, 2003). In a way, this philosophical challenge is not new and comes in many forms. It is similar to the one mentioned by the Greek historian, Plutarch (c. 46–120) who questions, in a thought experiment, whether a ship which is restored by replacing every one of its wooden parts remains the same ship.

In this way, if the genetic modification of the existing early embryo does not result in any substantial alteration, it would likely be regarded as analogous to somatic gene therapy, where the original identity of the individual remains unchanged. However, if the gene editing substantially alters the genome of a very early embryo, this raises more complex questions about whether the original embryonic individual continues to exist. Indeed, the genetic modification could effectively terminate the life of the initial embryo while generating a new embryo. Such a process would then involve selecting one living embryo over another based on the quality of its genetic heritage (MacKellar, 2017, 2023).

Gene editing of sperm, eggs, and during fertilization

Alternatively, if a genetic modification occurs either in the sperm and egg cells prior to their use or during fertilization in the formation of a single-cell embryo, this leads to the creation of a new individual who would not have otherwise existed. This would happen because even the slightest alteration in the factors contributing to conception results in a distinct person coming into existence in time who would not otherwise have existed. This expresses the important non-identity argument developed by the British philosopher Derek Parfit (1942–2017) (Parfit, 1976). In other words, the person who comes into existence because of any changes which happened before conception is non-identical (is a different person) to the one who would otherwise have existed had the changes not occurred.

Thus, once again, this process reflects a clear selective aspect, since it involves bringing a new individual into existence in preference to another who may, for example, have had a genetic disorder (MacKellar, 2017, 2023). Rather than being a form of *therapy*, which would involve treating an existing person for a condition, this approach ensures that only a specific individual is *selected* to come into being based on the characteristics of his or her genome. This means that it cannot be considered as having any clinical effect in the treatment of existing patients (MacKellar, 2019; Rulli, 2019).[3] It would also be unusual for the definition of *therapy* to be pushed to the extreme to include *selection* between different kinds of possible future persons.[4]

The use of heritable genome editing

With heritable genome editing the new genetic modification would not only affect the individual being generated but would be inherited by this individual's descendants, raising significant concerns about irreversible harm and the potential for the procedure to be used for the selection of future children.

Theoretically, it could be argued that heritable genome editing would be helpful in the very rare cases where some couples could not, otherwise, have a healthy child who is genetically related to them. For example, this may happen for prospective parents who both have dominant Huntington's disease. However, in such cases, the possibility of two affected persons wanting to have a child together will hardly ever occur unless they are brought together by the disorder itself (Baylis, 2019, p 31).

Alternatively, it could be suggested that heritable genome editing may help couples who are carriers of a genetic disorder, and who could have a genetically healthy and related child using preimplantation genetic diagnosis (PGD), but who may object to it on moral grounds, as it often involves the destruction of the non-selected embryos. However, this kind of argument relies on a misunderstanding, as heritable genome editing is not really an alternative to PGD because embryo selection remains a necessary step in any embryo editing procedure. This is because even after heritable genome editing is used to create embryos for reproduction (especially when dealing with complex genetic conditions), a risk remains of unintended consequences or incomplete edits, making the testing and eventual selection of the embryos a vital safeguard before they are used (Andorno et al, 2020).

Therefore, a thorough discussion is necessary to evaluate the effects on society arising from the selection of possible future children in this way. Modifications which change the genome of a human being, and all his or her possible descendants, raise significant ethical challenges (Visscher et al, 2025). This is also because such modifications may be inherently eugenic in nature. The word *eugenics*, which derives from two Greek roots '*eu*' (good) and '*genesis*' (birth), describes selection strategies or decisions aimed at affecting, in ways which are considered to be positive, the genetic heritage of a child, a community, or humanity in general (MacKellar and Bechtel, 2014, p 3). It was the Englishman Sir Francis Galton (1822–1911) who first coined the term *eugenics* in 1883 as he sought to implement into human beings selection procedures for inherited characteristics which had already been used, with success, in animal breeding programmes (Galton, 1883, p 25, n.1).[5]

However, many concerns exist relating to the consequences for society when it becomes possible to decide what kinds of children should be brought into existence through selective eugenic procedures (MacKellar and Bechtel, 2014). For example, a number of risks are associated to heritable genome editing which may undermine some important aspects of civilized society (Andorno et al, 2020).

Even from a purely safety perspective, it is important to recognize that inserting or deleting specific genes in the correct location within the genome of a developing embryo, without disrupting the biological equilibrium of the cells, is challenging and not without risk. A certain gene may influence a number of different characteristics, meaning that even if a gene was modified to influence a certain dysfunction this could give rise to unexpected consequences. The overall result would be a modification that may be more harmful than beneficial (President's Council on Bioethics, 2003, pp 38–39).

Developments in heritable genome editing

An important, though controversial, milestone in gene editing took place with the announcement, in 2015, that Chinese scientists had become the first to use the procedure to genetically modify a human embryo for research purposes (Liang et al, 2015). Partly in response to such groundbreaking research, the US National Academy of Science and National Academy of Medicine organized an International Summit on Human Gene Editing in 2015, in collaboration with the UK Royal Society and the Chinese Academy of Sciences. At the end of the summit, the organizing committee published a formal statement entitled *On Human Gene Editing*. This indicated, with respect to modifications using 'heritable or germline human genome editing' that:

> It would be irresponsible to proceed with any clinical use of germline editing unless and until:
>
> (i) the relevant safety and efficacy issues have been resolved, based on an appropriate understanding and balancing of risks, potential benefits, and alternatives, and
> (ii) there is broad societal consensus about the appropriateness of the proposed application. (Committee on Science, Technology, and Law, 2016)

Furthermore, in order to address the requirement for a broad societal consensus, the statement called for an ongoing international forum including a wide range of participants 'to establish norms concerning acceptable uses of human germline editing and to harmonize regulations, in order to discourage unacceptable activities while advancing human health and welfare' (Committee on Science, Technology, and Law, 2016).

Immediately following the summit, the two US National Academies formed an expert committee with a focus on science and law to prepare a comprehensive report examining the scientific perspective as well as the clinical, ethical, legal, and social implications of genome editing procedures in biomedical research and medicine. In 2017, this committee published the report entitled *Human Genome Editing Science, Ethics and Governance.* This included a number of suggested principles for human genome editing, namely promoting wellbeing, transparency, due care, responsible science, respect for persons, fairness, and transnational cooperation (Baylis, 2019, pp 133–134). It concluded that:

> Heritable germline genome-editing trials must be approached with caution, but caution does not mean they must be prohibited. If the technical challenges are overcome and potential benefits are reasonable in light of the risks, clinical trials could be initiated, limited to only the most compelling circumstances and subject to a comprehensive oversight framework that would protect the research subjects and their descendants; and have sufficient safeguards in place to protect against inappropriate expansion to uses that are less compelling or less well understood. (Committee on Human Gene Editing, 2017, p 189)

Significantly, in this new report, the conditions for heritable genome editing research had changed from being irresponsible 'unless and until' it is widely endorsed by society, to being permissible provided that certain criteria are satisfied (Baylis, 2019, pp 134–135). Moreover, in listing the necessary conditions for heritable genome editing to be applied to the fertility clinic, the 2017 report of the US National Academies no longer emphasized the 2015 call for developing broad societal consensus. According to the Canadian bioethicist Françoise Baylis, the aim for this was to (Baylis, 2019, pp 134–135):

a. Supplant any kind of non-expert democratic dialogue concerning whether sufficiently good reasons existed to consider heritable genome editing;
b. Emphasize that the crucial questions were not 'will it happen?' or 'should it happen?' but rather 'when will it happen?' and 'how should it happen?';
c. Accentuate the 'when' by proposing a ready-made framework which was generally based on self-regulation.

In other words, according to Baylis, this 2017 report sought to focus public attention on the challenges of 'how' to effectively move the science forward and that the ethical conditions were being considered in the light of making this happen (Baylis, 2019, pp 134–135).

In July 2018, the Working Group from the UK Nuffield Council on Bioethics published another report entitled *Genome Editing and Human Reproduction*, which in a similar way to the US National Academies report, indicated:

> We can, indeed, envisage circumstances in which heritable genome editing interventions should be permitted … we do not believe that there are absolute ethical objections that would rule them out in all circumstances, for all time. If this is the case, there are moral reasons to continue with the present lines of research and to secure the conditions under which heritable genome interventions would be permissible. (Nuffield Council on Bioethics, 2018, p 154)

Within this context, the report suggested that the procedure may be acceptable, but only in certain circumstances if, for example, no alternative procedure exists to avoid generating persons with genetic disease or who have a predisposition to such a disease (Nuffield Council on Bioethics, 2018, p 96). The Working Group also argued that heritable genome editing should only be considered when the following is taken into account:

- [W]here the procedure is carried out in a manner and for a purpose that is intended to secure the welfare of … a person who may be born as a consequence of treatment using those cells.
- [O]nly in circumstances in which it cannot reasonably be expected to produce or exacerbate social division or the unmitigated marginalization or disadvantage of groups within society. (Nuffield Council on Bioethics, 2018, p 157)

At the same time, an ever-growing number of academic articles were published suggesting that such procedures should not be ruled out in human reproduction if certain preconditions were met. For example, the Australian philosopher and bioethicist, Julian Savulescu and others, indicated that: 'We ought to use powerful technologies like Heritable Genome Editing to correct (…) inequalities and promote human flourishing. Such actions are moral imperatives' (Gyngell et al, 2019, p 522).

Accordingly, many commentators believe that heritable genome editing may be acceptable once safety and efficacy aspects have been addressed. A moratorium was then suggested by a number of individuals, national organizations, and other groups who believe that a permanent ban may not be the right way forward.

A further development relating to heritable genome editing took place in November 2018 when Chinese scientist He Jiankui announced that he had created the first human babies using the procedure (Greely, 2019). He did this by editing the genome of embryos using the CRISPR/Cas9 complex to remove a certain gene in order to confer genetic resistance to HIV (Kuo, 2018). The first babies were twin girls born in October 2018, with a third baby being born in 2019 (Begley and Joseph, 2018; Gutierrez, 2022).

However, the international community reacted with strong criticism and alarm to the announcement, and He Jiankui's work eventually received widespread international condemnation. A number of leading genome editing scientists shared their concerns about his experiment indicating that there was 'an inadequate medical indication, a poorly designed study protocol, a failure to meet ethical standards for protecting the welfare of research subjects, and a lack of transparency in the development, review, and conduct of the clinical procedures' (Baltimore, 2018, np). Significant concern also existed for the girls' wellbeing (Cyranoski and Ledford, 2018).

Because of this scandal, Chinese authorities eventually suspended He Jiankui's research activities and, in 2019, he was found guilty of forging documents as well as unethical conduct and sentenced to three years in prison (Yee and Hollingsworth, 2019).[6] But many questions remain about why He Jiankui undertook this experiment. For instance, in the absence of a formal moratorium, and in the light of influential reports such as the 2017 US National Academies report indicating that heritable genome editing should not be prohibited, He Jiankui may have believed that a certain amount of flexibility or even laxity existed. In other words, he could become a pioneer undertaking a contentious procedure which would eventually be accepted (Hurlbut, 2024, p 228). In this way, He Jiankui's controversial work may not represent a significant deviation from some of the debatable expectations of international science. It is possible that he may even have been moulded by them and, to some extent, have given expression to them. Accordingly, he may have believed that his experiment would increase his standing in the international scientific community, support China's reputation for pioneering research, while driving scientific progress forward by challenging ethical conservatism and societal concerns (Hurlbut, 2024, pp 220–226).

This means that presenting He Jiankui as a solitary rogue scientists may do injustice to a more complex reality. For example, questions may be asked relating to the role of the self-described 'scientific community' which seems to assert that it has sole authority to determine what should or should not be acceptable (Hurlbut, 2024, pp 225–231). The American historian of science, Benjamin Hurlbut, argues that self-governing science may eventually become rogue science if it ignores society's priorities and norms, explaining that: 'It is rogue not because it has abandoned its commitment to serving the good of society. It is rogue because it has appropriated from society the authority to declare what is good, dismissing ethical dissent as an ephemeral effect of a lagging society' (Hurlbut, 2024, p 231). Hulburt then explains that: 'He's experiment gave expression to an ethos in science that celebrates risk-taking, "disruptive" innovation, and the imperative to race ahead in the name of progress' (Hurlbut, 2024, p 220). But this may be considered inappropriate in a democracy where it is important that the whole of society considers the ethical challenges of heritable genome editing.

Is human heritable genome editing compatible with equality?

This study specifically examines *heritable* genome editing in the light of *equality* and does not discuss other ethical arguments relating to the procedure which may already have been discussed.[7] Indeed, based on the context outlined in the preceding pages, it is essential to explore whether human selection through heritable genetic modifications, which would enable the creation of certain kinds of individuals, can be compatible with the concept of equality.

In rational terms, it appears that, for human equality to even exist, something which is specifically *human* must be present in all human individuals to an equal extent.[8] For the English philosopher John Locke (1632–1704), this was obvious when he wrote:

> [T]here being nothing more evident than that creatures of the same species and rank, promiscuously born to all the same advantages of Nature, and the use of the same faculties, should also be equal one amongst another, without subordination or subjection. (Locke, 1689, Chap. II, Para. 4)

From a more substantial perspective, the concept of equality can be based on multiple factors such as: (1) human nature, as understood in philosophical terms; (2) shared DNA; (3) inherent value or dignity; (4) common origins, meaning similar ancestry; and (5) rationality. Of course, these different foundations are not mutually exclusive but can coexist. As the American social scientist John Evans states:

> Are humans those who have speech, reason, or some other trait? This is the philosopher's approach. Are humans those who are made in the image of God? That is the theologians' approach. Are humans those who have a typical human DNA sequence? That is the biologist' approach. (Evans, 2016, p 1)

In this study, the concept of equality will generally be defined in the same way as the 1948 UN's Universal Declaration of Human Rights, which indicates that:

- [R]ecognition of the inherent dignity and of the equal and inalienable rights of all members of the human family is the foundation of freedom, justice, and peace in the world (Preamble).
- All human beings are born free and equal in dignity and rights (Article 1)
- All are equal before the law and are entitled without any discrimination to equal protection of the law. All are entitled to equal protection against any discrimination in violation of this Declaration and against any incitement to such discrimination (Article 7).

However, some chapter authors will be developing their own discussion relating to other concepts of equality. This will include the distributive theory of equality which focuses on the fair allocation of resources, wealth, and opportunities within a society while seeking to prevent their concentration in the hands of a select few.[9]

In discussing inequalities, it is also possible to examine whether some form of imbalance may exist between the way persons are generated (MacKellar and Bechtel, 2014, pp 143–148). Indeed, it has been suggested that selection procedures for particular traits may limit a child from 'maximizing his chances for self-fulfilment' (Feinberg, 1980, p 135). The argument also contends that imposition of preferences by parents effectively makes them architects or bio-designers of their child's future. This means that, in exercising such control, parents may foreclose the possibility of an 'open future' thereby constraining their child's autonomy to determine the course of his or her own life. The child is then subjected to the genetic design of a third party, representing an unacceptable and unequal form of intergenerational control of parents upon their child (Kass, 2002, p 162).

Some form of inequality may also exist when certain persons have positional advantages (such as wealth), which they can use in generating children with certain positive traits through heritable genome editing. A tension may then exist between the positive characteristics that certain individuals may obtain from the procedure (giving them positional goods) in comparison to other individuals. Indeed, this may then challenge the inherent equality mentioned in the UN Declaration which is an absolute good with an intrinsic worth that is present in all.[10]

Moreover, it has been suggested that any form of selection of possible future children may have consequences on the way some existing children, born with a genetic disability, may be considered. Accordingly, the risks of inappropriate discrimination, such as ableism and stigmatization may be present when selection exists relating to specific traits. In part, this may also happen because no agreed understanding exists of what is meant by a severe genetic disease (Baylis, 2019). For example, some individuals of the Deaf community are concerned that the number of persons for whom this trait is genetic may decrease through deselective reproductive procedures.

They believe that they may then experience increased discrimination. Baylis explains:

> Diversity is an important social equalizer insofar as it makes it less possible to "Other" those who are different. If you homogenize a population, by "correcting" difference, you increasingly marginalize those who are outliers and can be readily identified as such. This marginalization is informed by the same values used to justify the initial selection of traits worth correcting. (Baylis, 2019, pp 68–69)

This all means that, if society wants to engage with the consequences of human selection such as with heritable genome editing, it will be necessary to examine the concept of equality and its implications.

Volume's outline

In order to address the previous questions, the book begins with a section examining how the concept of equality developed through history and philosophy as well as its applications in international law.

In the opening chapter, Matthew James contextualizes equality to establish a common understanding of the term before exploring its origins. In this way, the chapter charts the history of the term and its evolution in order to better comprehend a contemporary understanding of equality. Recognizing how the conversation relating to equality is still 'live', consideration is also given to its association with the concept of fraternity, which brings persons into a place that unites and then 'holds them' as they open up important questions of equality.

The chapter by Roberto Andorno continues along the same line but places greater emphasis on the philosophical reasons that have been presented throughout history in support of the concept of equality. He points out that questions around equality relate primarily to *values* rather than to *factual* measurable aspects of human beings, and that overlooking this reality has been a source of serious misunderstandings throughout history. The chapter stresses the insufficiency of merely utilitarian or contractarian arguments to provide a solid justification of equality. In this regard, Andorno considers that only the notion of human dignity can serve as a foundational principle of equality. Finally, the chapter considers whether, or to what extent, the principle of equality can be applied to relationships with future persons, which is a crucial question in the discussion on human heritable genome editing.

David A. Jones' chapter then analyses the evolution of the eugenic ideology since Francis Galton until the middle of the 20th century. This is because the Galtonian concept of eugenics, influenced by Darwinian ideas, suggests that natural selection mechanisms apply to humans and argues that inherent inequalities among individuals and groups justify intentional selective processes. After describing some precedents of eugenic ideas in Ancient Greece, the chapter examines whether eugenics, as a moral or political philosophy, is primarily focused on denying the equality of human beings or, instead, on emphasizing the ways in which they are unequal.

In her chapter, Britta van Beers examines how different understandings of equality in the human rights debate on heritable genome editing coexist. She indicates that the two main shapes of the principle of equality within the human rights governance of biogenetics are: (1) equality as the same intrinsic worth shared by all individuals as protection against genetic discrimination, eugenic practices and the genetic 'design' of children, and (2) equality as equitable access to the fruits of biogenetic science, which reflects a more 'distributive theory of equality'. In order to do justice to the first dimension of equality and to find a better balance between both understandings of this notion, she suggests bringing the principle of human dignity back into focus.

Laura E. Jacobs' chapter looks into the notion of the human genome as the 'heritage of humanity', as it appears in some international documents, such as the 1997 Universal Declaration on the Human Genome and Human Rights. Though acknowledging that this concept has evolved over time, she argues that one central underlying element has always been the notion of equality, which encompasses three different understandings: equal access, equality among generations, and equal dignity. This means that, depending on the context in which the heritage concept is applied, different forms of the notion of equality may be important. Based on this egalitarian nature of the notion, the heritage status of the human genome is then examined in the light of current legal instruments opposing heritable genome editing.

The volume continues, in a second part, by asking whether heritable genome editing may undermine equality in society. In other words, the book questions whether selecting future children, especially through heritable genome editing, should be accepted in human beings, while enabling new angles and perspectives to be carefully assessed.

In this way, Pia Matthews asks the question whether a 'life unworthy of life' actually exists in the case of very severe genetic disability. This is undertaken while discussing many factors, of which the most significant are notions of equality as well as injustice, discrimination, perceptions of what constitutes severe disability, attitudes to disability, ableism, the perception that death is preferable to disability, and (not least) anthropological assumptions about persons.

This is followed by a chapter from Calum MacKellar, who revisits the expressivist argument, which was previously explored in relation to some selective reproductive procedures, by extending it to heritable genome editing. This argues that the choice to use selective technologies to prevent the birth of individuals with specific disabilities may express disvalue towards existing people with those disabilities. This is undertaken by determining whether certain actions, decisions, or statements may express a discriminatory value system, just as clearly as if this had been revealed more directly through an oral or written declaration.

Whilst recognizing the potential benefits of heritable genome editing to address somatic disorders, Sandy Sufian and Rosemarie Garland-Thomson sound a more cautious note when it comes to considering the procedure. As disability studies scholars and women with genetic differences, they share their grave concerns about these 'genetic scissors' being used to effectively cut people like them out of existence. This is because they argue that they are whole beings and that their genetic conditions form a fundamental part of who they are.

Taking inspiration from the 1997 science fiction film *Gattaca*, Richard Playford sets out to respond to the question of whether heritable genome editing can be used without undermining equality. In doing this, he identifies a useful concept of equality that would be both applicable and action-guiding regarding genomic engineering. Playford then suggests how a plausible guidance may be provided to allow selection to take place whilst prohibiting more drastic possibilities which pose the risk of promoting inequality, discrimination, and undermining human dignity.

The last part of the book examines heritable genome editing and the possible future for society. Concerns exist that in democratic countries, it will be individual prospective parents, and not the state, who are likely to make decisions about whether to use such a procedure as they select which genetic traits they want for their future children. In his influential book, *The Future of Human Nature,* German philosopher Jürgen Habermas has labelled this trend as constitutive of what he terms 'liberal eugenics' (Habermas, 2003). However, it is worth noting that this parental choice will not be made in a vacuum since it may be influenced by societal norms, the technical options proposed by scientists, and possible prejudices concerning certain genetic traits (Baylis, 2019). For example, Baylis argues: 'The cumulative effect of these ostensibly personal reproductive choices will make plain our prejudices about what "perfect" and "imperfect" mean, in turn both revealing and determining who we are, and will be, as human beings' (Baylis, 2019, p 72). It has also been suggested that it is likely that the decision to use heritable genome editing will begin with initial condemnation, but that this will be followed by ambivalence, questioning, and limited use, which will then progressively change public perceptions to end up with widespread acceptance (Baylis, 2019, p 72).

Trevor Stammers focuses on the issue of equality of access. He argues that it is necessary to be realistic about aspirations of access to heritable genome editing, should the procedure ever become permitted and effective. Stammers highlights that both equality and inequality of access to heritable genome editing originate not with the arrival of a perfected technique but much earlier, with the consideration of position papers, opinion polls, and public consultations.

In his chapter, Vicente Bellver examines the concept of intergenerational equality while arguing that the integrity of future generations is crucial when considering the permissibility of heritable human genome editing. He notes that the idea of having responsibilities toward future generations has already been emphasized in various international documents focusing on environmental protection. In this context, he examines whether the application of the principle of intergenerational justice to the preservation of the human species may imply that an individual should not alter the genetic make-up of his or her descendants. This is because by modifying their genes, the individual may impose his or her limited perspective of the world onto future generations. Confronting what he describes as a 'gnostic' view that disregards nature, Bellver then studies whether international human rights instruments should aim to protect both the human species and the natural environment.

In the following chapter, Katie Hasson examines the impact of heritable genome editing on social equality, emphasizing the importance of context including social structures and power dynamics. She advocates for the involvement of civil society in discussions surrounding this technology, highlighting the absence of social justice perspectives in current major official discourses. The chapter outlines the risks associated with heritable genome editing, particularly its potential to exacerbate existing inequalities and undermine universal human rights, by enabling genetic modifications that could lead to a form of market-based techno-eugenics.

The next chapter by Benjamin Hurlbut explains that, especially since the Second World War and the development of human rights as an international project, the concept of equality is taken as a moral fact, a sort of foundation upon which elaborations of human integrity, dignity, and rights are constructed. He then examines the societal discussions over heritable genome editing in order to understand how the notion of equality is situated in relation to conceptions of the human person, the desire for meliorist genetic intervention, and most importantly, to rightly ordered deliberation and governance.

Notes

1 Procedures which were (and still are sometimes) considered use enzymes (very small biological machines) called nucleases that cut DNA. Some of these procedures are called zinc finger nucleases (ZFNs) and transcription activator-like effector nucleases (TALENs).

2 CRISPR is short for clustered regularly interspaced short palindromic repeats. Cas9 is short for Protein 9, which is an enzyme that breaks down DNA.

3 In addition, intentional heritable genome editing procedures cannot be seen as preventative treatments since such treatments only seek to avert and avoid, in a proactive way, the development of disease in persons who already *exist* (but not in non-existing persons).

4 This can be considered, instead, as a form of eugenics (which will be defined later).

5 It is worth noting that Galton did not mention coercive procedures when he defined the term eugenics.

6 For a background-study and summary of the He Jiankui scandal, see Kirksey, 2021.

7 Several volumes have already been published which include discussions relating to heritable genome editing in the light of a number of ethical arguments. See for example, Wilkinson, 2010; Agar, 2004; Galton, 2002.

8 This book will not address the possibility of nonhuman persons.

9 For a brief discussion of this see, European Parliamentary Research Service, 2022, pp 11–12.

10 This is also developed in MacKellar and Bechtel, 2014, pp 143–148.

References

Agar, N. (2004) *Liberal Eugenics: In Defence of Human Enhancement*, Oxford: Blackwell.

Andorno, R., et al (2020) 'Geneva statement on heritable human genome editing: the need for course correction', *Trends in Biotechnology*, 38(4): 351–354.

Baltimore, D. (2018) 'Statement by the Organizing Committee of the Second International Summit on Human Genome Editing', National Academies, 28 November 2018, Available from: www.nationalacademies.org/news/2018/11/statement-by-the-organizing-committee-of-the-second-international-summit-on-human-genome-editing [Accessed 10 January 2023].

Baylis, F. (2019) *Altered Inheritance*, Cambridge, MA: Harvard University Press.

Begley, S. and Joseph, A. (2018) 'The CRISPR Shocker: How Genome-Editing Scientist He Jiankui Rose from Obscurity to Stun the World', *Stat News*, 17 December 2018.

British Medical Association (2007) *Boosting Your Brainpower: Ethical Aspects of Cognitive Enhancements*, Available from: www.bma.org.uk/images/Boosting_brainpower_tcm41-147266.pdf [Accessed 27 August 2010].

Committee on Human Gene Editing (2017) *Scientific, Medical, and Ethical Considerations; Human Genome Editing: Science, Ethics, and Governance*, National Academies of Sciences, Engineering, and Medicine, Washington, DC: The National Academies Press.

Committee on Science, Technology, and Law (2016) *International Summit on Human Gene Editing: A Global Discussion: MEETING IN BRIEF.* Policy and Global Affairs; National Academies of Sciences, Engineering, and Medicine; S. Olson (ed), Washington, DC: National Academies Press; 1 January 2016, Available from: www.ncbi.nlm.nih.gov/books/NBK343651/ [Accessed 11 December 2024].

Cyranoski, D. and Ledford, H. (2018) 'How the genome-edited babies revelation will affect research – Some scientists worry the startling claim will lead to knee-jerk regulations and damage the public's trust in gene editing', *Nature*, 27 November 2018.

Dimitrievska, M., et al (2024) 'Revolutionising healing: Gene editing's breakthrough against sickle cell disease', *Blood Reviews*, 65: 101185.

European Parliamentary Research Service, Scientific Foresight Unit (STOA) (2022) *Genome Editing in Humans: A Survey of Law, Regulation and Governance Principles*, PE 729.506, Available from: www.europarl.europa.eu/RegData/etudes/STUD/2022/729506/EPRS_STU(2022)729506_EN.pdf [Accessed 4 June 2025].

Evans, J.H. (2016) *What is a Human?*, Oxford: Oxford University Press.

Evans, J.H. (2020) *The Human Gene Editing Debate*, Oxford: Oxford University Press.

Feinberg J. (1980) 'The child's right to an open future', in W. Aiken (ed), *Whose Child? Children's Rights, Parental Authority, and State Power*, Totowa, New Jersey: Rowman & Littlefield.

Galton D.J. (2002) *Eugenics: The Future of the Human Life in the 21st Century*, London: Abacus.

Galton, F. (1883) *Inquiries into Human Faculty and Its Development*, London: Macmillan.

Greely, H.T. (2019) 'CRISPR'd babies: human germline genome editing in the 'He Jiankui affair'', *Journal of Law and the Biosciences*, 6(1): 111–183.

Gutierrez C.N. (2022) 'What's next for the gene-edited children from CRISPR trial in China?, *New Scientist*, 29 June 2022.

Gyngell, C., Bowman-Smart, H., and Savulescu, J. (2019) 'Moral reasons to edit the human genome: picking up from the Nuffield report', *Journal of Medical Ethics*, 45(8): 514–523.

Habermas, J. (2003) *The Future of Human Nature*, Cambridge: Polity.

Harris J. (2007) *Enhancing Evolution*, Princeton: Princeton University Press.

Hurlbut, J.B. (2024) 'Advancing progressively backwards: Guiding and governing heritable genome editing' in N. Baer (ed), *The Promise and Peril of CRISPR*, Baltimore: Maryland, Johns Hopkins University Press, pp 219–244.

Kass, L.R. (2002) *Life, Liberty and the Defense of Dignity: The Challenge for Bioethics*, San Francisco: Encounter Books.

Kirksey, E. (2021) *The Mutant Project: Inside the Global Race to Genetically Modify Humans*, Bristol: Bristol University Press.

Kuo, L. (2018) 'China orders inquiry into "world's first gene-edited babies"', *The Guardian*, 27 November 2018.

Li, H., et al (2020) 'Applications of genome editing technology in the targeted therapy of human diseases: mechanisms, advances and prospects', *Signal Transduction and Targeted Therapy*, 5(1): 1.

Liang P., et al (2015) 'CRISPR/Cas9-mediated gene editing in human tripronuclear zygotes', *Protein Cell*, 6(5): 363–362.

Locke, J. (1689) *Two Treatises of Government, Concerning the True Original Extent and End of Civil Government*, Available from: www.yorku.ca/comninel/courses/3025pdf/Locke.pdf [Accessed 11 December 2024].

MacKellar, C. (2017) 'The gene editing of human embryos and the new eugenics', *Bioethica Forum*, 10(2): 74–76.

MacKellar, C. (2019) 'Genome modifying reproductive procedures and their effects on numerical identity', *The New Bioethics*, 25(2): 121–136.

MacKellar, C. (2023) 'Bioethical decision making and genome editing', in E. Valdés and J.A. Lecaros (eds), *Handbook of Bioethical Decision Making: Volume 1*, Switzerland: Cham, Springer Nature, pp 29–42.

MacKellar, C. and Bechtel, C. (2014) *The Ethics of the New Eugenics*, New York: Berghahn Books.

National Academies of Sciences, Engineering, and Medicine (2017) *Human Genome Editing: Science, Ethics, and Governance*, Washington, DC: The National Academies Press.

Nuffield Council on Bioethics (2018) *Genome Editing and Human Reproduction: Social and Ethical Issues*, London: Nuffield Council on Bioethics.

Ossorio P.N. (2003) 'Inheritable genetic modifications' in A.R. Chapman and M.S. Frankel (eds), *Designing Our Descendants,* Baltimore: Johns Hopkins University Press, pp 259–262.

Parfit, D. (1976) 'Rights, interests and possible people' in S. Gorovitz et al (eds) *Moral Problems in Medicine.* Englewood Cliffs: Prentice Hall, pp 369–365.

President's Council on Bioethics (2003) *Beyond Therapy: Biotechnology and the Pursuit of Happiness*, Washington, DC: The President's Council on Bioethics.

Rulli, T. (2019) 'Reproductive CRISPR does not cure disease', *Bioethics*, 33: 1072–1082.

Sas, D.F. and Lawrenz, H.M. (2017) 'CRISPR-Cas9: The latest fashion in designer babies', *Ethics & Medicine*, 33(2): 81–95.

Sparrow R. (2022) 'Human germline genome editing: On the nature of our reasons to genome edit', *The American Journal of Bioethics*, 22(9): 4–15.

Visscher, P.M., et al (2025) 'Heritable polygenic editing: the next frontier in genomic medicine?' *Nature,* https://doi.org/10.1038/s41586-024-08300-4.

Wilkinson, S. (2010) *Choosing Tomorrow's Children: The Ethics of Selective Reproduction*, Oxford: Oxford University Press.

Yee, I. and Hollingsworth, J. (2019) 'Chinese scientist who edited genes of twin babies is jailed for 3 years', *CNN News*, 30 December 2019, Available from: https://edition.cnn.com/2019/12/30/china/gene-scientist-china-intl-hnk/index.html [Accessed 11 December 2024].

Zhang, Yr., Yin, Tl., and Zhou, Lq. (2023) 'CRISPR/Cas9 technology: Applications in oocytes and early embryos', *Journal of Translational Medicine*, 21(1): 746.

PART I

Equality: Its Origins and Applications in International Law

1

The Origins of Equality

Matthew James

Introduction

Writing at the turn of the new millennium, the leading biotech entrepreneur Gregory Stock made the salient comment that 'Enhanced humans will manifest and reinforce their philosophy in their biology' (Stock, 2002, p 123). It is widely accepted that each generation has the power to influence the next generation in a myriad of ways. In particular, parents have the opportunity to shape the lives of their children, ranging from insisting that they study hard with additional tutor support outside of school, to ensuring they engage in a range of extracurricular sporting or performing arts activities. In the vast majority of cases, this comes down to a seemingly innate desire of all parents to give their children the best possible start in life. To enable them to go beyond what they, as parents, were able to achieve and accomplish. The advent of the biotech century and key advances in the field of genetics greatly enhances this power of influence that progenitors could exercise over their progeny. The extent to which a child's life could be moulded and shaped could become increasingly irresistible, something which did not escape the US President's Council on Bioethics. In its 2003 report, *Beyond Therapy: Biotechnology and the Pursuit of Happiness,* it noted 'in as much as the consequences of genetic screening are imposed before birth and are carried as the child's permanent biological destiny, the inegalitarian effect of the new technology is unprecedented and irreversible' (2003, p 52).

Over 20 years since the Council's report was published, the challenges are no less significant. As our understanding of the human genome has expanded so have techniques to edit and manipulate our genetic code. Making changes to the human genome that can be heritable to future generations is now a real possibility and with it the very scope for injustices to be perpetrated by one generation on another through the misuse of its powers. If we have responsibilities to future generations at all then they are not simply to prevent certain abuses, but also to see how best we can use our capacities for the greater good. The challenge to this generation is to develop our science and philosophy such that human beings are empowered to flourish freely and not to control or dominate. A plethora of ethical and social questions come to the fore, coalescing around key themes of justice, fairness, and equality. But how are we to understand equality in this new biotech age? Can it truly exist in whatever 'brave new world' lies ahead of us? Can we define equality and how it has evolved?

The focus of this chapter is to set equality in context, offering some introductory comments to establish some common understanding of the term before attempting to plot something of its history and origins. In many respects, it will assume the nature of a history lesson, charting the origins of the term and its evolution to better understand what has shaped and influenced contemporary thinking. In addition, the chapter will identify how the conversation is still alive with ideas as to what we conceive both equality and fraternity to be and what this could mean for the future.

Defining equality

As in the days of 18th-century France when the streets echoed with the cries of the revolutionists proclaiming 'Liberty – Equality – Fraternity', equality and debates over perceived challenges to equality may never be too far from the public conscience. From news reports to political debates, economic analysis to discussions within the workplace of equality, diversity, and inclusion, equality seems to be one of the defining characteristics of modern talk. Assuming an etymological perspective initially, 'equal' is derived from the Latin *aequus* meaning 'even, level, equal'. Applying this to human interaction and basic anthropological implications as to what it means to be human, to be on an 'even, level' footing, 'equal' to one another appears to reflect a fundamental belief reflected in international declarations such as the United Nations' Universal Declaration of Human Rights (1948). Nevertheless, there is a clear distinction between aspirational intent and the practical outworking of this. As British novelist and poet, George Orwell reflected in his allegorical novella, *Animal Farm*, 'All animals are equal, but some animals are more equal than others' (1987, p 90), criticizing the hypocrisy and corruption of the pigs who take over the farm. Replace animals for humans and the quote loses nothing of its potency in describing the challenges faced in contemporary society.

How we define equality is therefore crucial to our understanding of the term. The terms 'equality', 'equal', and 'equally' signify a qualitative relationship. 'Equality' speaks of 'correspondence between a group of different objects, persons, processes or circumstances that have the same qualities on *at least one respect*, but not all respects' (Stamford Encyclopaedia of Philosophy 2021). There is scope for difference and variance, but there remains one specific feature which unites them all and places them on a *level* playing field of interaction. Perspective is gained by contrasting equality with identicality which could be mistakenly assumed as a synonym of equality but contrasts far more distinctly. Identity has a specificity that equality does not, referring to one and the same object in relation to itself in *all of its* features.

Furthermore, both the terms equality and identicality need to be distinguished from the term 'similarity'. From the Latin root *similis*, meaning 'like', to saying that something is similar is to recognize a state of being that is 'like, alike; resembling but not the same; of the same kind or amount' (Oxford Mini Dictionary, 1992, p 486). Therefore, to say that all humans are equal is not to say that they are identical but there are similar possessing qualities common to all but which are not exactly the same. Crucially, this helps to make a distinction which can often be missed in contemporary debates. When we come to make a judgement about equality, we are presuming there is at least some difference between the two things being compared.

Equality's evolution

Given its place in understanding the workings and interaction of humanity, in terms of relationships, status, rights and opportunities, much has been written and explored in terms of equality's relevance and importance as both a philosophical, economic, and social term. Taking a step back, a different picture emerges.

There appears to be a paucity of material which explores and discusses, in depth, the history of equality and how it has evolved; something which only in recent times is being remedied and reflected in the literature. In broad terms, something of what is recognized today as equality can be traced back to the time of the Ancients. Placing a stake in the ground metaphorically at this point unfurls a timeline that charts times of war, conflict, social and political revolution, social and racial justice all of which lead to the recognition of the emergence of equality as an indicator of 'modern society' from the 18th century onwards. It achieved prominence in the second half of the 20th century embodied in the post-World War II United Nations Charter proclaiming the goal of securing the 'equal rights of men and women' throughout the world regardless of 'race, sex, religion, or language' (UN General Assembly, 1948).

It could be said that equality forms the moral horizon of humanity. We may not be able to readily identify how it rose in its importance, but it has nevertheless become a significant part of our ethics and moral reasoning, advanced by our models of education and culture making. Philosopher Jeremy Waldron comments on this in terms of the aim of policy development:

> In the voluminous modern literature on egalitarianism, there is a tremendous amount on equality as a policy *aim*. Philosophers ask whether we should be aiming for *equality* of wealth, *equality* of income, *equality* of happiness, or *equality* of opportunity; they ask whether equality is an acceptable aim in itself or code for something else, like the mitigation of poverty; they ask whether aiming for equality implies an unacceptable levelling; whether, if achieved, it could possibly be stable; how it is related to other social values such as efficiency, liberty and the rule of law; and so on. (Waldron, 2002, pp 1–2, emphasis mine)

From the time of the Ancients to the early 21st century, theoretical reflection on equality has remained strong despite the seeming lack of explicit work exploring the evolution and existence of this seemingly foundational concept. Rather like the waves of the sea flowing back and forth, yet it always remains. It is not a mere aspiration or aim but what Waldron has referred to as a 'background commitment' (Waldron, 2002, p 2), but regardless of whether it is background or foreground in our thinking, trying to understand its history helps to elucidate its meaning and relevance.

Talk of equality triggers associations with rights and equality before the law, which leads back to the work of the 'Father of liberalism' (Korab-Karpowicz, 2010, p 291), namely John Locke. Reviewing his work shows references to equality as having roots in the Christian faith, in particular his theory of property proffering a reading of the biblical doctrine that God gave the earth to *all* humanity, *res communis,* with individuals acquiring property in this communal inheritance through working the land (Locke, 1986, Ch 5, Sec. 27). While one can look back to the past and identify equality's existence, any sense of historical context which *explains* its evolution is hard to come by. From a historical perspective it appears to only confirm what could be coined as equality's 'orphan status'. Historian Jonathan Israel reflects that:

> Neither the philosophical nor the historical grounding of this idea [equality], that is its intellectual origins and roots, is at all obvious and this whole issue has been, to a quite remarkable extent, shrouded in neglect in the historical academic literature. (Israel, 2006, p 545)

Similarly, historian David Armitage laments the fact that, 'there is still no history of how the movements against all the major heritable forms of domination and subordination – monarchy, aristocracy, slavery, and gender differences – intersected with or diverged from one another [...] And the conceptual history of equality remains almost entirely unwritten' (Armitage, 2014, p xx). Some put this difference down to labelling talk of equality as bourgeois ideology and false consciousness, which in terms of Marxist and Engelian thought would thwart the realization of a communist revolution (McMahon, 2019, p 115).

The closer we try to get to equality, to identify its origins, to understand and recognize it, the more it becomes blurred and obscure. Comparisons could be made with the term human dignity which has also proven particularly vexing to clearly define within the field of bioethics.

Equality as a word is readily recognized and used from campaign slogans to media soundbites, activist speeches to court verdicts, yet the term has become detached from experience. The French sociologist, Pierre Rosanvallon, comments: 'We face a crisis of equality … It has become a sort of remote deity which is routinely worshipped but has ceased to inspire any living faith' (Rosanvallon, 2013, p 8). Very often the immediate context for consideration of equality has fallen on exploring what could be termed 'material equality' and its relevance to economics. The recent work of Thomas Piketty in *A Brief History of Equality* (2023) is but one example of such an exploration. With some acknowledgement of the philosophical context within which to explore equality, the book largely focuses on a history of equality through the lens of economics and finance: in essence what could be termed 'material equality'. The Dutch Historian, Siep Stuurman's voluminous exploration of humanity, *The Invention of Humanity: Equality and Cultural Difference in World History* (2017), makes a valuable contribution to discussions by taking some serious steps to document equality's history. While acknowledging the value of material equality, in chartering the history of equality, Stuurman chooses not to pursue this in any depth but seeks to define what he describes as cross-cultural equality or 'common humanity'.

This is the notion enshrined in the United Nations Universal Declaration of Human Rights, which refers to the 'recognition of the inherent dignity and of the equal and inalienable rights of all members of the human family' (Preamble). The use of the term 'recognition' intends to express that the notions of dignity and equality were not capriciously 'invented' by policy makers, but rather 'discovered' by them. Equality came to be recognized with no little effort by men and women in societies in which tribalism, ethnocentrism, and stark inequalities were the empirical norm. The premise that Stuurman appears to be advocating is that only when human beings can imagine circumstances other than their lived realities can they begin to do something about them and overcome them. With subtle tones more readily associated with a self-help guru: 'If you can think it and see it, you can have it,' Stuurman comments, 'Bringing into existence social dreams and imagined futures, equality thoughts create their own reality' (2017, p 6). It is why he writes of equality's invention, rather than discovery.

To help make sense of the equality terrain, Stuurman uses three meta concepts:

1. Common humanity – what all human beings share.
2. Anthropological turn – the ability to recognize the logic and difference of other cultures.

3. Time – temporality and history which locate people on developmental scales or in relation to how we create and order the world.

These three work together to make sense of cross-cultural equality. He takes each one separately for the purposes of discussion and analysis, to compare and contrast, noting similarities and difference. But overall, there is a unity of all three that helps to complete the picture. With these three concepts in place to help with the tracing and mapping of the terrain, Stuurman attempts to chart the evolution of equality which has been eagerly sought after by many. Drawing upon Stuurman's expansive work, a condensed survey of the historiography of equality will be presented and discussed here.

Axial Age

To borrow from German-Swiss philosopher Karl Jaspers, we start at the Axial Age (c. 8th century BC). This was the period generally recognized as the time when the great intellectual, philosophical, and religious systems that came to shape human society and culture emerged. The Ancients held to an organic and hierarchical conception of the world, one that was essentially antiegalitarian. All of nature, including the social world, consists of a series of interlocking entities, each with its own built-in goal. Each entity in turn was a means to the satisfaction of a higher-level goal.

The social world was highly stratified to reflect differences of ability that in turn led to differences of function and a corresponding difference in status. A collective good is seen to be evident: a good that was more important than, and subsumed, all the lesser goods. This consisted of the survival of the city as an internally self-ruled entity. It was the city, or polis, that was the locus of freedom, understood as self-rule. Freedom was not something understood or worked out on the individual level. Instead, individuals were fulfilled when they performed their relevant proper function in maintaining the city's freedom.

The writing of the great political thinkers of the time reflected these sentiments. The eminent Greek philosopher Plato's work *Republic*, in which we see set out the idea of a just society identified as a harmonious society, in which the division of labour was exactly correlated with individual differences of ability (1987, 422e and 423b-d). Similarly, for Aristotle, shared status was the basis: 'Equality consists in the same treatment of similar persons' (1952, VII, 14). Believing that eudaimonia and the well-lived life was achieved through the practice of virtue, the Stoics took Aristotelian thinking and formulated the position that *everyone* was deserving of the same dignity and respect, placing emphasis on the natural equality of all rational beings.

Building upon this, the teaching and beliefs of early New Testament Christianity followed a similar approach, establishing the basis of what would be recognized as a modern conception of equality. Christianity proclaimed the equal moral worth of persons in the eyes of God. It is the special respect in which all human beings are in fact equal, drawing upon both Stoic doctrine and Hebrew thought from the book of Genesis that all human beings 'male and female' were created in the 'image of God' (Genesis 1:27). This doctrine of equality is further expounded by the apostle Paul (c. 5–c. 64/65 AD), a Christian leader and teacher in the early Church, who wrote that 'There is neither Jew nor Greek, there is neither bond nor free, there is neither male nor female: for you are all one in Christ Jesus' (Galatians 3:26–29. See also Colossians 3:10–11). This idea is also upheld in the Jewish Talmud and reflected in Islamic teaching although not as explicitly. While no mention of the 'image of God' is made in the Quran, a statement which resonates with the verses in Genesis is found in hadith, describing how God created Adam in his image (*bi ṣūratihi*).

What we can take away from these beliefs are two key ideas. First that recognizing equality among human beings requires that we treat them in a special way. Second, that this 'special treatment' could either entail ensuring every human is treated in the same way or that they are assisted in different ways in order to restore them to, or to aid them in reaching, the specific factual state (Capaldi, 2002, p 4). Furthermore, despite what may appear to be (in the first instance) different philosophical and theological schools of thought, there is a common thread running through early Judaism, Confucianism, Greek thought and Stoicism, Christianity and Islam – that of seeing culturally 'the other' as part of a common human family. Initially this appears to provide the basis upon which to achieve world peace and harmony and yet such a harmonious outcome is never realized. Instead, what separates (as opposed to what unites), becomes the focus, and a great divide opens up between those who choose to believe these tenets of faith and those who do not. As such, in both the cases of Christianity and Islam, faith beliefs served as effective tools to help build and justify an empire against the perceived 'barbarians' (Stuurman, 2017, pp 68–69; 84–85).

This gives rise to the emergence of what Stuurman calls the anthropological turn, comprising of formations of ideas and language which express human similarities based on networks of trade, exchange and conquest, rather than anything grounded in the ideas of religion and philosophy. This is exemplified through the Greek historian and geographer Herodotus' encounters with the non-Greek barbarians of the ancient world, the Persians and Scythians. In what he frames as 'David and Goliath' style encounters, Greece fights off the threat of external rule and develops on its own. This in turn leads to the levels of creativity, innovation, and other intellectual achievements we have come to recognize from the Greeks. This common story of conquest and empire expansion is seen in the accounts of Sima Qian (c. 145 BC–c. 86 BC), the Herodotus of classical China, who undertook the first-ever universal history of China and its neighbours, seeking to understand the complexity of empire. Similarly, in the European encounters with the New World such as those involving the work of French Renaissance philosopher Michel de Montaigne (1533–1592) and others. They tried to understand and appreciate the equality and common humanity as witnessed through different peoples and cultures, as opposed to conquering and displacing them which tended to be the default *modus operandi* of the time.

Modernity

Moving into the era of modernity, equality became a central notion because of the Protestant Reformation. Once more arising from division and separation, in this case a separation from the Roman Catholic Church due to differences of opinion on matters of doctrine, Protestantism was a religious reform movement that swept through Europe in the 1500s. Many of the attributes and values that have come to be recognized today as 'modern values', such as democracy, scepticism, capitalism, individualism, civil rights, and equality, arose from the fervour and zeal of the Protestant Reformation.

The philosopher Nicholas Capaldi differentiates between the classical viewpoint and the modern viewpoint based on an understanding of standards. From a classical perspective, all standards of truth, goodness or beauty were based on structural features of the world which were external to human beings. Through either knowledge or grace, some would have authority to be able to access such standards and conform to them which was seen as wisdom (Capaldi, 2002, p 5). In contrast, for the modern thinker the focus fell on internal standards and laying hold of them would likely lead to an encounter of a transcendent and/or external order. This understanding is reflected in many areas of life including the fields of science and art. Copernicus' work shifted the focus away from humans being at the centre of the universe (as per Ptolemy's geocentric model of the world) and challenged the idea of humans occupying a place of privileged observers of the universe. This relativity of perception principle saw human observation as representative of observations from the average position in the universe. Similarly, the work of Renaissance artists displayed real emotions, elevating the exploration of perspective and allowing the viewer to connect with them.

Modernity essentially transformed how humans related to the world around them. As Capaldi observes, 'The apprehension of internal standards required that we conform to them, but conformity to internal standards came to mean the transformation of the external world to conform to these internal standards' (Capaldi, 2002, p 6). The hierarchical order of the Ancient World of organic associations was now rejected and replaced with a new idea of nature as a *mechanism* created by God. As humans we reflect and replicate God's creativity by transforming our world, inspired and guided by an internally apprehended divine vision. This idea of the inner-directed individual was encapsulated in what became known as the 'Protestant work ethic' with its focus on work and achievement. One of the four principles of the Reformation *sola scriptura* ('by scripture alone') affirmed the belief that it was only the Word of God that directs and binds the conscience of the Christian, establishing the notion of Christian liberty. This liberty upheld equality before the law, with differentiation based on achievement thereby permitting a meritocracy.

Reflecting the general tenure of the Protestant worldview, English philosophers Thomas Hobbes and John Locke developed doctrines of natural law and social contract theory. In Hobbes's eyes, in their natural condition, individuals possess equal rights, because over time they have the same capacity to do each other harm (1651). Locke argued that all human beings have the same natural right to both (self-) ownership and freedom (1690, IV, Sec. 22–23).

Enlightenment

The 17th and 18th centuries witnessed an intellectual movement sweep through Europe. Known as the Enlightenment, much fervour drove an animated exchange of new ideas and thought resulting in a cornucopia of fresh ideas and thinking concerning God, reason, nature, and humanity. These 'enlightened ideas' sought to make sense of the human predicament based on scientific, rational thought, as a response to the thinking of religion and traditional authority which had preceded it. It is perhaps not surprising that the Enlightenment marks a significant shift from the ideas of the Axial Age to the ideas of modern equality. Reason and abstraction becoming the new touchstone, replacing religious revelation and the divine.

One of the central philosophers of the Enlightenment period, Immanuel Kant (1724–1804) was influenced by Protestant thinking. His moral philosophy centred around the idea of the categorical imperative which helped to formulate the belief of equality being grounded in universal human worth. For Kant, the principle of human rights was informed by his thinking on autonomy and self-legislation upon which the freedom of all rational beings was founded (Allison, 2001, pp 118–119). These enlightened ideas helped initiate the great modern social movements and revolutions and were taken up in modern constitutions and declarations of human rights. The French Revolution, one of the most notable due to the way in which it upended the feudal system, absolute monarchy, and transformed the landscape of France's politics through the rise of nationalism, saw equality, along with freedom and fraternity, become a basis of the 'Declaration of the Rights of the Man and of the Citizen' of 1789 (*Déclaration des droits de l'homme et du citoyen* of 1789). Fashioned by revolutionary flare and zeal, this understanding of equality holds that, while recognizing differences, human beings are to be regarded as equals. This is often referred to as 'human equality' or 'basic equality' or 'equal worth' or 'human dignity' (Kateb, 2014; Waldron, 2017; Rosen, 2018). As Waldron notes, whether these terms are synonyms is a matter of interpretation, but 'they cluster together to form a powerful body of principle' (Waldron, 2017, p 3).

From this brief historical sweep of equality's evolution, it would be easy to conclude that equality gradually emerges triumphant and takes hold through civilization. As Stuurman suggests, this is true to an extent but there is more of a tussle involved as equality seeks to stake its position. It develops an essentially contested nature. As has been established, equality is regarded as a natural human condition based on shared qualities of faculties, reason, language or bodily needs. Thus, if equality is regarded as natural, then inequality opens up a conversation challenging and critiquing the social foundations, structures and dynamics of the day through the lenses of gender, race, religion, and rank. This saw a profound change in the debate: 'equality received the benefit of the doubt while inequality had to be justified by rational argument' (Stuurman, 2017, p 259). At the same time that the Enlightenment provided the tools and the ideas with which to contest and critique a form of hierarchy and control, it also had the power to establish new forms of social control and structures which only served to reconstitute inequalities in new ways. The conception of modern equality could therefore be considered as Janus-faced (Stuurman, 2017, p 259). Taking inspiration from the Roman God, Janus, the god of beginnings, transitions, and endings, Stuurman considers modern equality in one direction, matched with modern inequality in the other. This was seen in at least four aspects:

(a) New expressions and conceptualization of political economy which help to justify social inequality on the basis of utility and productivity
(b) New theories of gender establishing difference between the sexes on the basis of nature.
(c) New theories of race which established natural hierarchies on the basis of the taxonomies of natural history.
(d) New philosophies and understanding of history which understood the story of humanity as unfolding through set stages of development.

Reflecting on these aspects in the context of world history, one can see how (b) and (c) in particular have had a significant impact on framing Western perceptions of cross cultural (in)equality. This has seemingly given license to Western civilizations to impose their ideas as superior and the model for which all others must follow. Consideration of the events of the beginning of the 20th century and various social movements concerning race, colour, and gender reveal how deeply entrenched this thinking has become. Hence, why Stuurman's use of the term 'Janus-faced equality' is astute.

At the same time, there has been some push back against such thinking, not least in terms of European efforts for social reform, as seen in the abolition of the slave trade spearheaded by Christians committed to putting their faith in action and seeing reform of society. Similarly, Frederick Douglass and Martin Luther King Jr, prominent activists, social reformers, and statemen of the 19th and 20th centuries, respectively, exemplified through their conduct and practice their adept ability to combine their Christian beliefs with more modern (Enlightenment-derived) understandings of equality. Consequently, Stuurman posits that:

> [P]erhaps we may conclude that Enlightenment modern equality strengthened and systematized Christian equality, while Christian evangelical equality heightened the emotional temper of modern equality and elevated common humanity to the status of a sacred ideal. (2017, p 385)

The elevation to a sacred ideal is perhaps no better illustrated than by the 1948 UN Declaration of Human Rights, in which modern equality was accepted as a global standard. Out of the ashes of the atrocities of Nazi Germany, the Declaration ushered in an era of human rights in the second half of the 20th century and helped to established modern equality as an inclusive concept within global society. Without adopting a particular set of philosophical or religious persuasions to define the nature of humanity, the Declaration's success was that it was 'couched in a new inclusive language of modern equality' (Stuurman, 2017, p 508).

Whatever success the Declaration can be said to have had in establishing modern equality as a global standard, the apparent human tendency to make sense of the world by majoring on difference as a key criterion continues to hold strong. Despite the growing interdependence of the world's economies, cultures, and populations, brought about by the cross-border exchange of people, services, money, and information, arising from globalization, tensions have not eased. It would appear equality cannot be easily uncoupled from those who are deemed unequal in some form, either implicitly or explicitly. Political equality seems to be founded upon an idea of a 'community of equals' that identifies itself as superior to other communities. This is perhaps best illustrated by Athens or Sparta, where the 'equal ones' or 'men of equal status' who possessed the political privileges of citizens, were defined explicitly against the unequal: *metics*, women and slaves, who represented the core of society. This was, in essence, what Aristotle was stating when he defined equality as the 'same treatment of similar persons': equality was only for equals.

In turning to gene editing, comparisons can be made to contemporary society, which immediately triggers thoughts of the kind of society portrayed in the 1997 science fiction film *Gattaca*, where genetic discrimination may be illegal, but eugenics is practiced by using genotype profiling to identify valids (the equivalent of the 'equal ones' of Athens perhaps) to qualify for professional employment while in-valids are relegated to menial jobs. Must modern equality inevitably rest on assertions – explicit or implicit – of inequality?

A range of modern commentators has been willing to broach such speculation. The German philosopher, Friedrich Nietzsche (1844–1900) was perhaps the boldest. 'The will to equality is the will to power' (1968, p 277). It was his view that where power was concerned there was always, in his view, domination, and hence inequality, of one kind or another. As writer and futurist Alvin Toffler has commented, 'if we do not learn from history, we shall be compelled to relive it. True. But if we do not change the future, we shall be compelled to endure it. And that could be worse' (1972). Heritable genome editing appears to hold tremendous power to correct inequalities at the same time as exacerbate them still further. History indicates that new technologies of this kind tend to be largely determined by those who have the power to do this. Oxford academic C.S. Lewis foresaw this in his strikingly prophetic vision of the 21st century:

> [W]hat we call Man's power over Nature turns out to be a power exercised by some men over other men with Nature as its instrument.... There neither is nor can be any simple increase of power on Man's side. Each new power won *by* Man is a power *over* Man as well. Each advance leaves him weaker as well as stronger. (2002, pp 420–421)

There is a certain resonance here with Stuurman's use of the phrase 'anthropological turn' and how we perceive change in relation to one another. Advances in heritable genome editing appear to present us with the opportunity to consider afresh what it means to be human and how we perceive and relate to others, against the all too familiar backdrop of seeking to arrive at a form of utopian society. Is there a place for equality in such a future?

French social theorist Jacques Attali has considered the impact and implications of utopian thought. As noted earlier in this chapter, globalization has led to a growing interdependence of the world's economies, cultures, and populations, which in turn has triggered a new type of 'nomadism' which Attali sees as characterized by a willingness of individuals to leave their country of origin to settle somewhere else which boosts their chances of success and opportunity (Attali, 1999). This pursuit is further fuelled by advances in science and technology with the promise of improved quality of life and success. From Attali's analysis, there are four ideas at the heart of what has been thought of as utopia: eternity, freedom, equality, and fraternity. He writes:

> At the dawn of societies, men saw their passage on Earth as nothing more than a labyrinth of pain, at the end of which stood a door leading, via their death, to the company of gods and to *Eternity*. With the Hebrews and then the Greeks, some men dared free themselves from theological demands and dream of an ideal City where *Liberty* would flourish. Others, noting the evolution of the market society, understood that the liberty of some would entail the alienation of others, and they sought *Equality*. (Attali, quoted in Joy, 2000)

Attali posits that these different utopian goals exist in tension in society. Establishing on an altruist foundation, the fourth utopia, *Fraternity*, links individual happiness with the happiness of others, affording the promise of self-sustainment. In considering the future of these four concepts, Attali believes equality is the more difficult to see as a possible solution. As we have seen from the survey of equality's origins earlier in this chapter, inequality has existed for so long that trying to reach a place of equality seems impossible. Attali therefore believes fraternity is far more achievable since we live in relationship and contact with one another, whether in terms of colleagues, family, or friends. Thinking in terms of our relationships with one another and extended notions of fraternity can help to influence and shape our understanding of economy, commerce, the environment and ecology, science and technology. Attali's ideas, here, would seem to help enrich our understanding of equality as opposed to merely replacing it, a perspective which is taken up by political philosopher, Antonio Maria Baggio, who underscores the stabilizing effect that both equality and fraternity can bring. While recognizing the fact that fraternity has its roots in the French Revolution, it did not seem to evolve to the same status of liberty and equality. Nevertheless, like a three-legged stool, there is a need for fraternity to help bring stability to liberty and equality. It remains central to the discourse and yet remains hidden, occupying a position of relative obscurity.

American Political philosopher John Rawls (1921–2002) appears to focus on justice and has very little to say about the place or value of fraternity. But scratch beneath the surface and it becomes apparent that he does work with an understanding of fraternity in his conceptions of civic friendship and servility. In his book *A Theory of Justice* he proposes the principle of difference which is built upon the idea of fraternity in terms of social cooperation and reciprocity in such a way that sees social and economic inequalities arranged so that the greatest benefit is realized by the least advantaged (1999, p 90). How Rawls conceived fraternity is what Baggio and others identify as the challenge surrounding fraternity's central yet obscure place in discussions. In the second half of the 19th century, political regimes have taken fraternity to justify dominion and the restriction of freedom of others (for instance, as seen in socialist regimes of Eastern Europe), or as a nationalistic bond that goes way beyond patriotism and devotion to one's country, manifesting as hate and rejection of the foreigner (Baggio, 2013, p 45). The point being made is that fraternity cannot be understood as a taxonomy of fraternities because in so doing it excludes and denies the universal dimension of fraternity. It speaks to what a human being is based, on a horizontal, not a vertical, form of relationship. Fraternity indicates a uniting cause, a relationship of parity between different subjects. Baggio writes:

> Fraternity carries with it a *principle of reality*, which explains the constitution of a human being. Each man or woman I meet even though he or she may not be my brother or sister is truly or potentially a brother or sister to someone else, and is for this reason a carrier of natural rights to liberty and equality which fraternity guards in unity. (Baggio, 2013, p 48)

Therefore, it appears that an important yet often unrecognized link between equality and fraternity may exist. A human condition which is both enlightening and complex, fraternity establishes the space in which all humans come together and from where questions of liberty and equality can be identified and addressed.

Concluding thoughts

Equality has a somewhat obscure status. It is a widely recognized term, clearly featuring on the moral horizon of humanity and a commitment upon which many other ideas and concepts are based. Nevertheless, understanding its evolution and what it expresses can be challenging due to what could be termed an 'orphan-like' status. Yet without the term in our lexicon, we would probably need to invent such a word. The implications of heritable genome editing present the opportunity to wield significant power and influence over the next generation which, in turn, poses questions of justice, fairness, and equality. We therefore need to consider the place and importance we give to equality in society and how we add strength and standing to the term.

Turning to consider fraternity as a 'partner' to equality may help in this regard. With its association of civic friendship and servility, fraternity possesses a similar covert status as equality. Yet working in relation to one another, a powerful synergy can be established. Fraternity helps to create a space in which there is recognition of the common connection across all of humanity. The advent of new biotechnologies can easily contest this connection as the pursuit of the individual and individuality often reigns supreme over the concerns and welfare of the group or community. Beginning with fraternity helps to bring us into a place that unites us, to then 'hold us' as we open up important new questions about equality in the age of genome editing.

References

Allison, H.E. (2001) 'Immanuel Kant 1724–1804', in T. Honderich (ed), *The Philosophers. Introducing Great Western Thinkers*, Oxford: Oxford University Press, pp 115–121.

Aristotle (1952) *Politics*, translated by Benjamin Jowett, Overland Park: Digireads.com.

Armitage, D. (2014) 'Foreword', in R.R. Palmer (ed), *The Age of the Democratic Revolution: A Political History of Europe and America, 1760–1800*, Princeton: Princeton University Press.

Attali, J. (1999) *Fraternités. Une nouvelle utopie*, Paris: Fayard.

Baggio, A.M. (2013) 'The forgotten principle: Fraternity in its public dimension', *Journal of Dialogue & Culture,* 2(2): 35–52.

Capaldi, N. (2002) 'The meaning of equality', in T.R. Machan (ed), *Liberty and Equality*, Stanford, CA: Hoover Press, pp 1–33.

Hobbes, T. (1651) in E. Curley (ed), 1994, *Leviathan*: With Selected Variants from the Latin Edition of 1668, Indianapolis: Hackett.

Israel, J. (2006) *Enlightenment Contested: Philosophy, Modernity, and the Emancipation of Man 1670–1752*, Oxford: Oxford University Press.

Joy, B. (2000) Why the future doesn't need us. Wired, Available from: www.wired.com/2000/04/joy-2/ [Accessed 4 November 2024].

Kateb, G. (2014) *Human Dignity*, Cambridge, MA: Harvard University Press.

Korab-Karpowicz, W.J. (2010) *A History of Political Philosophy: From Thucydides to Locke*, New York: Global Scholarly Publications.

Lewis, C.S. (2002) 'The abolition of man', in *C.S. Lewis Selected Books*, London: HarperCollins.

Locke, J. (1986) *The Second Treatise of Government* (first published in Two Treatises of Government [1690]), Buffalo, NY: Prometheus Books.

McMahon, D.M. (2019) 'To write the history of equality', *History and Theory*, 58(1): 112–125.

Nietzsche, F. (1968) *The Will to Power*, translated by Walter Kaufmann and R.J. Hollingdale, New York: Vintage Books.

Orwell, G. (1987) *Animal Farm*, London: Penguin Books.

Oxford Mini Dictionary (1992) *Oxford Mini Dictionary*, Oxford: Oxford University Press.

Piketty, T. (2023) *A Brief History of Equality*, Cambridge, MA: Harvard University Press.

Plato (1987) *The Republic*, London: Penguin Classics.

President's Council on Bioethics (2003) Beyond therapy: Biotechnology and the pursuit of happiness, Available from: https://bioethicsarchive.georgetown.edu/pcbe/reports/beyondtherapy/index.html

Rawls, J.B. (1999) *A Theory of Justice*, rev edn, Cambridge, MA: Harvard University Press.

Rosanvallon, P. (2013) *The Society of Equals*, translated by Arthur Goldhammer, Cambridge, MA: Harvard University Press.

Rosen, G. (2018) *Dignity: Its History and Meaning*, Cambridge, MA: Harvard University Press.

Stamford Encyclopedia of Philosophy (2021) Equality. Stamford Encyclopedia of Philosophy, Available from: https://plato.stanford.edu/entries/equality [Accessed 1 November 2024].

Stock, G. (2002) *Redesigning Humans: Our Inevitable Genetic Future*, London: Profile Books.

Stuurman, S. (2017) *The Invention of Humanity. Equality and Cultural Difference in World History*, Cambridge, MA: Harvard University Press.

Toffler, A. (1972) *The Futurists*, New York: Random House.

UN General Assembly, Universal Declaration of Human Rights, 217 A (III), 10 December 1948, Available from: www.refworld.org/legal/resolution/unga/1948/en/11563 [Accessed 4 November 2024].

Waldron, J. (2002) *God, Locke, and Equality: Christian Foundations in Locke's Political Thought*, Cambridge, UK: Cambridge University Press.

Waldron, J. (2017) *One Another's Equals: The Basis of Human Equality*, Cambridge, MA: The Belknap Press of Harvard University Press.

2

What Makes Human Beings Morally Equal?

Roberto Andorno

Introduction

The notion of moral equality among humans has been explored by philosophers throughout history. In Western Antiquity, the idea of universal moral equality was more of an exception than a norm in highly hierarchical societies that accepted slavery and strictly differentiated individuals based on their social status. During this period, the concept of equality developed, particularly among two groups: the Stoic philosophers, such as Cicero and Marcus Aurelius, and Christian theologians and philosophers, including Augustine and Boethius. The former emphasized that all human beings are naturally equal due to their shared rational nature. From this shared rationality, Stoic philosophers concluded that there is a universal moral law, known as 'natural law', which all people can comprehend through reason. Christian thinkers also appealed to the common rational (that is, spiritual) nature that distinguishes human beings from other forms of life, but they also added a theological argument: the belief that every human soul is created 'in the image and likeness of God', which makes them equal before God and grants them the same dignity.[1]

However, it was not until the 19th and 20th centuries that the principle of universal moral equality, advocated by the philosophers of the Enlightenment, was politically recognized and formally enshrined in national constitutions and international human rights instruments. This process reached a global milestone in 1948 with the adoption of the Universal Declaration of Human Rights, which states that '[a]ll human beings are born free and equal in dignity and rights' (Article 1).

Certainly, there have also been scholars in modern times who critically viewed the principle of equality, but they were quite rare. The German philosopher Friedrich Nietzsche is perhaps the most prominent advocate of the inegalitarian position. In his works, he repeatedly argued that individuals are not equal, but fundamentally different. Therefore, any efforts to establish a society based on equality would ultimately result in tyranny and the suppression of individuality. In *Beyond Good and Evil* (1909), Nietzsche criticized what he called 'slave morality', which promotes equality and which he associated with Christian morality.

In his book *Thus Spoke Zarathustra* (1920), Nietzsche introduced the concept of the Superman (*Übermensch*), which society should strive to promote (§ 5). This figure represents the next stage in human evolution, a being who transcends traditional morality and religion, creating his own values. Zarathustra, a kind of prophet who is the central character of the book, contrasts the *Übermensch* with the mediocre 'last man' of modern egalitarian societies. The Superman symbolizes the human being of the future, who possesses extraordinary strength of will, intellect, and creativity.

When writing the introduction to *Thus Spoke Zarathustra,* Nietzsche's sister, Elisabeth Förster-Nietzsche, quotes him as saying: 'With the help of favourable measures, great individuals might be reared who would be both different from and higher than those who heretofore have owed their existence to mere chance. Here we may still be hopeful: in the rearing of exceptional men' (p x).

Considering this and other similar passages, it is not surprising that Nietzsche's Superman has often been associated with ideas like those seen in eugenics programmes from the first half of the 20th century, notably the Nazi programme. However, this question has been the subject of much debate among scholars. Some argue that Nietzsche's emphasis on individual autonomy and creativity is incompatible with such programmes. Others claim that, regardless of such theoretical speculations, Nietzsche's call to 'breed the Superman' directly influenced racist and eugenics movements in Europe prior to World War II, including in England (Stone, 2002).

The fact is that a few decades after Nietzsche, the Nazi regime exploited his inegalitarian ideas for its own racist purposes and implemented them with the dramatic consequences that are well-known. There is no need to mention that today, most philosophers consider Nietzsche's inegalitarian views to be deplorable and unacceptable.

This chapter first examines three major arguments in support of the principle of moral equality, a fundamental pillar of modern democratic societies. It then considers if, or to what extent, this principle can be applied to the relationship between generations, given that this is one of the important issues at stake in the discussion around heritable genome editing in humans.

When entering this discussion, it is important not to mistake equality as a *descriptive* notion with equality as a *moral* notion, as this confusion has led to significant misunderstandings in the past. In this chapter, equality is viewed as a moral (or normative) concept. Thus, the question 'Are human beings equal?' relates primarily to *values* rather than to a *factual,* measurable aspect of human beings. People are obviously unequal in various aspects, such as intelligence, physical appearance, strength, intellectual and physical abilities, health condition, and socio-economic status. However, this chapter does not focus on these factual inequalities. Instead, it is dedicated to exploring equality as a *value.* Ultimately, moral equality means that all human beings possess a common rank by virtue of which they have a claim to equal respect and consideration.

The fundamental question that arises here is: 'What makes humans equal?' Is equality grounded on purely utilitarian considerations or on social agreement? Or is there any deeper reason behind this basic value of social life? It is also important to consider how, or to what extent, the principle of equality, which was traditionally applied to existing individuals, can be relevant to our relationship with future persons. This question is especially important in the debate on human heritable genome editing, as this technique can potentially have a negative impact on individuals who do not yet exist and future generations in general. It is noteworthy that the procedure can be used to serve eugenic purposes and 'design' children according to the parents' wishes and ideals.

At this point, two preliminary clarifications regarding the supposed 'benefits' of human germline gene editing are necessary. Firstly, it is conceptually misleading to claim that this procedure serves 'therapeutic' purposes. This technology does not cure or prevent diseases in any existing individual. Instead, it aims to *create* children with certain genetic features that some may consider advantageous from a health perspective. This fact makes this procedure categorically different from somatic gene therapies (Rulli, 2019; Andorno et al, 2020). Secondly, the resort to human germline engineering cannot be justified by invoking the supposed benefit for all of humankind. The goal of removing harmful genes from the entire human population or adding new genes to improve the human species (such as extending lifespan or enhancing physical or mental abilities) is more utopian than realistically achievable.

Therefore, the crucial question is: Can the principle of equality provide a basis for the obligation to preserve the integrity and identity of future generations of children?

Equality: a foundational notion of modern societies

The US Declaration of Independence (1776) famously presents the notion that 'all men are created equal' as a 'self-evident truth'. Similarly, the French 'Declaration of the Rights of Man and the Citizen' of 1789 begins by proclaiming emphatically that 'men are born and remain free and equal in rights'.[2] In the 20th century, the 1948 Universal Declaration of Human Rights (UDHR) solemnly recognizes 'the inherent dignity and of the equal and inalienable rights of all members of the human family' as 'the foundation of freedom, justice and peace in the world' (Preamble). Article 1 of the UDHR states that '[a]ll human beings are born free and *equal* in dignity and rights', mirroring the wording of the 1789 French Declaration. This provision may seem like a merely rhetorical statement, but it is actually revolutionary in human history. As the American human rights and public health advocate Jonathan Mann pointed out, Article 1 of the Declaration marks a 'seismic shift in human consciousness, and is so profound that, paradoxically, its importance may not be fully realized' (Mann, 1998). Indeed, for the first time in history, and following the horrors of the Second World War, the international community was able to reach a consensus on a legal-political instrument stating that all human beings have equal and inherent dignity and rights.

The fact is that, today, the entire international human rights system that emerged from World War II is explicitly grounded on the moral conviction that every human being deserves to be treated as an *equal* to other human beings. As Jeremy Waldron puts it, today there is a common political understanding that we are all 'one another's equal' (2017).

The legal philosopher Ronald Dworkin emphasized the same idea when he advocated for a principle he called 'equality of concern and respect' (1977, p 180). His analysis highlights what some scholars refer to as an 'egalitarian plateau', which indicates that the principle of equality provides a common ground accommodating diverse ethical and political philosophies, such as utilitarianism and libertarianism (Kymlicka, 1990, p 5; Brown, 2009, pp 83–101). As a matter of fact, today, all major theories of political philosophy share a fundamental commitment to equality, considering that the lives of every single individual matter and matter equally. Certainly, these theories may differ on the best ways to secure the satisfaction of basic human interests. Some of those approaches focus on individual liberty, while others emphasize the importance of social justice. However, all of them agree that each person is of equal intrinsic value, regardless of age, health condition, ethnic origin, socio-economic status, gender, or any other circumstance. Today, most political theories agree that how we treat people stems from the fundamental moral assumption of moral equality. Nonetheless, this assumption, which is enshrined in national Constitutions and international legal instruments, raises a fundamental question regarding the grounding of moral equality.

Three possible foundations for moral equality

Why do we generally see the principle of moral equality among human beings as true? Is this assumption simply a rhetorical statement, or is it founded on valid reasons? Among the possible grounds for moral equality, the following three can be mentioned:

(a) A pragmatic foundation

A possible answer to the above question is that moral equality is primarily based on *pragmatic (or empirical)* reasons. This kind of argument can be found, for instance, in the UDHR, which refers to the importance of recognizing equal dignity and rights for all individuals to prevent 'barbarous acts which have outraged the conscience of mankind' from occurring again (Preamble). According to this approach, the recognition of moral equality would be the result of historical experiences demonstrating that, if this principle is ignored, it leads to dehumanizing and cruel practices in society.

Therefore, the requirement of equal respect would result from the observation of the inherent vulnerability of human beings. The word 'vulnerability' itself comes from the Latin *vulnus*, which means 'wound'. Vulnerability relates to the human capability of being wounded, either physically or mentally. From this perspective, *all human beings* are vulnerable because they are all susceptible to being harmed by others. In human rights discussions, the term 'vulnerability' refers more specifically to the increased susceptibility of certain individuals or groups to be harmed or mistreated by others or by the state. Populations that are particularly at risk of being harmed, exploited, or discriminated against include children, women, older adults, people with disabilities, and members of ethnic or religious minority groups.

Sociologist Bryan Turner takes this pragmatic-empirical perspective a step further when he suggests that *vulnerability is the very foundation of universal human rights*. In his view, human rights exist because 'human beings share a common ontology grounded in a shared vulnerability' (2006, p 6). He further explains that, aware of their inherent fragility, humans feel the need to establish social institutions and governmental power to protect themselves from violence and misuse. The unprecedented devastation caused by the two world wars of the 20th century has heightened an awareness of human vulnerability and the fragility of the institutions designed to protect individuals from violence and exploitation and led to the establishment of the current international human rights system.

This pragmatic approach is certainly right in pointing out that the awareness of our common vulnerability has been – and still is – a crucial incentive in the efforts aiming at the promotion of human rights. It is indisputable that the modern international human rights system emerged as a response to the horrors of the Second World War. However, we cannot draw from this historical circumstance the conclusion that human rights are *morally* grounded in vulnerability. Vulnerability is a mere fact and a feature of the human condition, not a value per se (Andorno, 2016). To establish moral equality, a *normative* foundation is necessary, as discussed below.

(b) A contractarian basis

Another possible justification for moral equality is the idea of a social contract as the basis of society. This approach suggests that equality stems simply from our collective decision to treat people equally. Thus, we are all regarded as morally equal simply because we have collectively reached a consensus on this important principle. In political philosophy, several authors have argued in this way throughout history. Among the classic social contract thinkers are the Englishmen Thomas Hobbes and John Locke, and the Genevan philosopher Jean-Jacques Rousseau. The latter famously claimed in his work *Du contrat social* (1762) that we are all equal by nature, and to ensure equal rights we have created the state by means of a social contract.

In modern times, John Rawls is perhaps the most well-known contractarian legal philosopher. He describes a hypothetical scenario called the 'original position', in which a group of individuals must come to an agreement about the type of political structure they want for society. They do this without knowing what role they will ultimately occupy within that society. This leads them to establish an egalitarian political structure in which they all have equal rights (Rawls, 1971).

From a similar perspective, it has been claimed that, after the adoption of the Universal Declaration of Human Rights in 1948, the issue of foundations of equal human rights has already been 'solved', and there is no need to continue searching for any objective or absolute ground (Bobbio, 1996). Since the vast majority of governments have officially adopted the Declaration, this instrument would prove that a value system can be based solely on consensus.

The contractarian position has the practical advantage of not requiring any further philosophical explanation. However, it is unsatisfactory from a deeper moral perspective. Does the mere factual circumstance of an egalitarian social contract really provide a *moral* basis for equality? Is this not a kind of circular argument, as it essentially claims that we are all equal before the law because the law states that we are all equal? Moreover, what if, hypothetically, a society decided to establish an unequal social structure that discriminates against certain groups? Let us not forget that modern history is replete with examples of inegalitarian societies. Notable instances include the apartheid political system in South Africa, which lasted until the early 1990s, and the officially sanctioned ethnic discriminatory practices in the United States prior to the civil rights movement in the 1960s.

A related question is whether equality is fair simply because we have agreed upon it or, on the contrary, if we have agreed on equality because we consider it to be fair. This second statement appears to better reflect the truth: The requirement of fair treatment must arise from some substantial reasons, and not merely from a formal agreement (which could have any content).

Another problem with social contract theories concerns our obligations to future generations. It is difficult to see how people who do not yet exist can be parties to our social contract. This difficulty arises mainly because there is no opportunity for mutually beneficial interaction between distant generations. Certainly, the quality of life of future generations (for instance, the preservation of the environment) depends largely on the decisions of the present generation. However, the decisions of future generations cannot retroactively affect our quality of life. Therefore, a purely contractarian view seems inadequate to respond to the need for intergenerational justice, not only regarding the protection of the environment but also concerning the preservation of the human genome against improper manipulations.

(c) Human dignity as a basis for equality

A deeper explanation for moral equality than the previous ones is the idea that every individual has inherent worth or *dignity.* According to this view, we are entitled to equal rights not just because we are vulnerable or because the social contract may uphold equality, but because we possess *inherent dignity.* This is the explanation given by international human rights law that was developed in the aftermath of the Second World War.

The UDHR emphasizes that human rights are grounded on the recognition of the 'inherent dignity and the equal and inalienable rights of all members of the human family', as stated in the Preamble. Consistently, Article 1 asserts that 'all human beings are born free and equal in dignity and rights'. Clearly, human dignity serves as the foundation for all human rights and freedoms, including equality. Even more explicitly, the 1966 International Covenants on Civil and Political Rights, and on Economic, Social and Cultural Rights, affirm in their Preambles that human rights '*derive* from the inherent dignity of the human person'. Hence, human dignity is not regarded by international law as a collective term for rights, but as the ultimate source of all rights.

What does the adjective 'inherent', as used in the UDHR, mean? According to the dictionary, this term means 'involved in the constitution, or essential character of something', 'intrinsic', or 'permanent or characteristic attribute of something' (Merriam-Webster Online Dictionary). When 'inherent' is employed to qualify human dignity, it highlights the idea that dignity is *inseparable from the human condition*. Human dignity is, therefore, not to be understood as an accidental quality of some human beings or as a value derived from some features of certain human beings, but as the worth that *all human beings possess by simply sharing in the human condition.*

A direct corollary of inherent dignity is the right to *equal* treatment and freedom from discrimination. Indeed, if human dignity is inseparable from the human condition, it is the same for all. Therefore, if dignity is the source of all human rights and the same for all, all human beings possess *equal* basic rights (Andorno, 2009, p 229).

In other words, people deserve respect *because* they possess inherent dignity. Therefore, it is conceptually incorrect to argue that the notion of dignity is 'useless' and can simply be replaced by 'respect', as the American bioethicist Ruth Macklin claims (2003). Although the concept of 'dignity' is very close to that of 'respect', both notions are not synonymous. Respect for persons is the *consequence* of human dignity, not dignity itself, just as a bell's sound is an effect produced by the bell, not the bell itself. In fact, the idea of dignity is not identical to that of respect simply because dignity provides the rationale for demanding respect for every human individual (Andorno, 2009).

It is true that defining the concept of dignity in very precise terms is a challenging task. However, this does not mean that dignity is merely a rhetorical or political notion without any substance. Despite the difficulty in clearly defining dignity, we must acknowledge that there are ways to better convey the meaning of this powerful concept. One of them is the famous Kantian requirement of non-instrumentalization of human beings. According to Kant's third formulation of his categorical imperative, we should always treat people as an *end in themselves*, and never merely as a *means* to our ends (1996, p 79). People should not be instrumentalized because they are not 'things', but 'persons' (that is, rational beings). While things have a 'price' and can therefore be replaced by an equivalent (like money), 'dignity' makes every person unique, precious and irreplaceable.

Another approach to understanding the concept of dignity is the so-called 'negative way', which consists in considering what opposes dignity rather than what aligns with it. The reason for this is that malevolence is easier to recognize than beneficence. Using this *via negativa* it has been argued that dignity conveys the idea that people must '*never* be treated in a way that denies the distinct importance of their own lives' (Dworkin, 1994, p 236), or that 'there are some things that should not be done to anybody, anywhere' (Midgley, 1999, p 160).

Not surprisingly, while most human rights are defined in positive terms by international law ('Everyone has the right to…'), those norms that aim to protect people from the most serious violations of human dignity are framed negatively, as prohibitions: 'No one shall be held in slavery or servitude'; 'no one shall be subjected to torture or to cruel, inhuman or degrading treatment or punishment' (Articles 4 and 5 of the UDHR, respectively). Precisely because these rights and freedoms are directly derived from human dignity they are thought of as *absolute*, in the sense that they are not subject to any exception.

In short, modern democratic societies view the requirement of equal respect for every single individual as a direct corollary of human dignity and as a fundamental prerequisite for the just functioning of the human community. As Rawls puts it, 'each person possesses an inviolability founded on justice that even the welfare of society as a whole cannot override' (1971, p 23).

In relation to this, British bioethical scholars Deryck Beyleveld and Roger Brownsword have famously proposed a distinction between 'dignity as empowerment' and 'dignity as constraint' (2002), which they present as 'rival notions'. While the former emphasizes freedom to pursue one's autonomously chosen goals, the latter is more concerned with promoting the common good of society and humankind. However, this distinction may be somewhat misleading, as it oversimplifies the issue. These two facets of human dignity do not really constitute rival notions. They are not mutually exclusive but complementary, in the same way that 'rights' and 'duties', or 'freedom' and 'responsibility', are not rival but complementary concepts (Andorno, 2009).

Does the principle of moral equality also apply to intergenerational relationships?

The prospect of human germline alterations prompts a question about the implications of the principle of moral equality. Does this principle also apply to intergenerational relationships? More specifically: do we bear a duty not to intentionally alter the genetic integrity and identity of future individuals? In other words, does the creation of genetically modified children violate the principle of moral equality?

The preliminary response to this latter question is yes, but with the caveat that equality holds a particular significance in this broader context of intergenerational relationships. Few authors have explored this matter in such depth and with greater engagement than the German–American philosopher Hans Jonas. Responsibility for humankind's future is the central tenet of his influential book *Imperative of Responsibility* (1984). He argues that, due to the immense technological powers we have gained over ourselves and our environment, we are now burdened with significant long-term responsibilities that accompany this power.

Jonas points out that while traditional ethics concentrated on interpersonal relationships – specifically, the interactions between coexisting individuals – the 'new ethics' he advocates deals with questions of intergenerational justice. According to him, the aim of this new 'ethics for the technological civilisation' is not only to avoid causing harm to our *neighbour*, but also to prevent harm to those who *do not yet exist*. It is our new responsibility to preserve the existence and identity of future generations, ensuring decent and humane living conditions for our descendants. To convey this notion,

Jonas proposes a new 'categorical imperative' for our technological society that runs:

> Act so that the effects of your action are compatible with the permanence of genuine human life. Or expressed negatively: Act so that the effects of your action are not destructive of the future possibility of such life. (Jonas, 1984, p 11)

Even more emphatically, Jonas argues that 'caring for the future of humankind is the overruling duty of collective human action in the age of a technological civilization' (1984, p 136). According to him, the present generation has a 'metaphysical responsibility' to preserve not only the natural environment but also the essential features of the *human condition* from a misuse of our increasing technological powers.

Accordingly, the present generation does not have the right to genetically design future children, not even with the best of intentions. Our self-understanding as free and autonomous human beings largely depends on being able to perceive ourselves as 'subjects', and not as a kind of 'living artifacts' designed by others. If children were genetically engineered to have certain skills and to excel in a particular field, they might feel immense pressure to conform to that genetic destiny, potentially limiting their free exploration of other interests and talents.

In discussions about obligations to children it is common to cite what Joel Feinberg called the child's 'right to an open future' (Feinberg, 1983). This notion is regarded as a guiding consideration in discussions of issues as diverse as testing children for late-onset genetic disorders, declining cochlear implants for deaf children, and many other decisions that may be contrary to children's best interests. The case of parents trying to modify possible future children's genome to make them closer to their ideal is explicitly mentioned in the literature as a violation of the children's right to an open future (Millum, 2014, p 522). The ultimate goal of this right is to enable children to make their own decisions (when possible) on issues that are crucial for their lives, and to preserve their future autonomy.

It is true that parents make many choices that shape their children's lives (for instance, education, lifestyle, etc.). However, intentionally selecting their children's heritable genetic information would be a step too far and would seriously exceed parental authority. Future individuals should continue having the freedom to develop their potential without being genetically conditioned by the prevailing ideas of 'good' and 'bad' traits and skills that were dominant at the time they were conceived. Moreover, it is noteworthy that even if germline gene alterations were originally meant to improve health and prevent diseases, it is very likely that these initial steps would gradually progress toward making genetic changes for non-therapeutic enhancements. According to several experts, this slippery slope is virtually unavoidable, since the line between treating illness and 'improving' human beings is blurred (Lanphier et al, 2015). The German philosopher Jürgen Habermas has also made a significant contribution to this debate. In his view, the use of germline gene editing in human reproduction would grant us disproportionate power over our descendants and result in a dramatic loss of liberty for future generations (2003). Of course, this loss of liberty should not be understood in the traditional sense as implying external restrictions on individuals' freedom, which may involve coercing people to do, or refrain from doing, certain things. What is at stake in germline gene editing is a much more dramatic, profound, and *internal* constraint *on* individuals' freedom. Let us not forget that our genome is the set of genetic instructions that determine or significantly influence most of the features that characterize each of us and make each of us a unique and irreplaceable individual. Intentionally selecting that information by someone else's decision introduces a seriously disturbing factor in the self-perception of the genetically engineered individual as an autonomous (that is, free) subject.

In Habermas' view, preventing the manipulation of our descendants' genome is crucial for preserving our capacity to see ourselves as the authors of our own life history (2003, p 23). Interestingly, we are only now realizing that the contingency surrounding natural procreation is not morally neutral but has intrinsic ethical value because freedom from external design is necessary for us to be able to express our true selves. To put it another way, the principle of equality between generations crucially depends on the condition that every person's features continue to be the result of *chance* rather than intentional *choice* by others.

These considerations are not purely academic but have already been enshrined in international documents and approved by the international community. For instance, the UNESCO Declaration on the Responsibility of the Present Generations Towards Future Generations, adopted in 1997, provides that:

> [T]he human genome, in full respect of the dignity of the human person and human rights, must be protected and biodiversity safeguarded. Scientific and technological progress should not in any way impair or compromise the preservation of the human and other species. (Article 6)

Accordingly, the 1997 Universal Declaration on the Human Genome and Human Rights provides that the human genome should be preserved as it is, 'in a symbolic sense, the heritage of humanity' (Article 1). This statement relates to the international law concept known as the 'common heritage of humanity', which asserts that certain highly valuable natural and cultural resources should be preserved in trust for future generations, protecting them from misuse and exploitation. Consistently with that principle, human germline alterations are labelled by the Declaration as 'practices that could be contrary to human dignity' (Article 24).

Conclusions

In summary, the answer to the question: 'What makes human beings morally equal?' is that their intrinsic dignity makes them equal. Neither merely pragmatic reasons nor any hypothetical social contract can by themselves make people equal. Only the recognition of people's inherent worth can provide a solid foundation for moral equality among individuals and serve as a barrier against discrimination and other forms of unfair treatment. The international human rights system is explicitly based on this understanding as it affirms that human rights '*derive* from the inherent dignity of the human person' (1966 International Covenants on Civil and Political Rights, and on Economic, Social and Cultural Rights, Preambles).

Furthermore, the principle of equality does not only apply to currently living individuals. It also underpins the obligation to protect the identity and integrity of future generations from a misuse of biotechnological developments such as germline engineering, which would lead to a radical commodification of future humans. Genetic knowledge should be used solely to promote the wellbeing of individuals and society, and not as a tool to impose intergenerational control.

Notes

1. See Book of Genesis 1:26. More specifically, the Letter of St Paul to the Galatians stresses moral equality by emphasizing that 'there is neither Jew nor Greek; there is neither slave nor free; nor is there male and female, for you are all one in Christ Jesus' (3:28).
2. 'Les hommes naissent et demeurent libres et égaux en droits'.

References

Andorno, R. (2009) 'Human dignity and human rights as a common ground for a global bioethics', *Journal of Medicine and Philosophy*, 34(3): 223–240.

Andorno, R. (2016) 'Is vulnerability the foundation of human rights?', in A. Masferrer and E. García Sánchez (eds), *Human Dignity of the Vulnerable in the Age of Rights*, Dordrecht: Springer, pp 257–272.

Andorno, R., Baylis, F., Darnovsky, M., et al (2020) 'Geneva Statement on Heritable Human Genome Editing: The need for nourse correction', *Trends in Biotechnology*, 38(2): 351–354.

Beyleveld, D. and Brownsword, R. (2002) *Human Dignity in Bioethics and Biolaw*, Oxford: Oxford University Press.

Bobbio, N. (1996) *The Age of Rights*, Cambridge: Polity Press.

Brown, A. (2009) *Dworkin's Theory of Equality*, London: Palgrave Macmillan.

Dworkin, R. (1977) *Taking Rights Seriously*, Cambridge, MA: Harvard University Press.

Dworkin, R. (1994) *Life's Dominion. An Argument about Abortion, Euthanasia and Individual Freedom*, New York: Vintage.

Feinberg, J. (1983) 'The child's right to an open future', in J. Howie (ed), *Ethical Principles for Social Policy*, Carbondale, IL: Southern Illinois University Press, pp 97–122.

Habermas, J. (2003) *The Future of Human Nature*, Cambridge: Polity Press.

Jonas, H. (1984) *The Imperative of Responsibility. In Search of an Ethics for the Technological Age*, Chicago: University of Chicago Press.

Kant, I. (1996) 'Groundwork of the metaphysics of morals', in M. Gregor (ed), *Practical Philosophy.* The Cambridge Edition of the Works of Immanuel Kant, Cambridge: Cambridge University Press, pp 37–108.

Kymlicka, W. (1990) *Contemporary Political Philosophy*, Oxford: Clarendon Press.

Lanphier, S., Urnov, F., et al (2015) 'Don't edit the human germline', *Nature*, 519: 410–411.

Macklin, R. (2003) 'Dignity is a useless concept: It means no more than respect for persons or their autonomy', *British Medical Journal*, 327: 1419–1420.

Mann, J. (1998) 'Dignity and health: The UDHR's revolutionary first article', *Health and Human Rights*, 3(2): 30–38.

Midgley, M. (1999) 'Towards an ethic of global responsibility', in T. Dunne and N.J. Wheeler (eds), *Human Rights in Global Politics*, Cambridge: Cambridge University Press, pp 160–174.

Millum, J. (2014) 'The foundation of the child's right to an open future', *Journal of Social Philos*ophy, 45(4): 522–538.

Nietzsche, F. (1909) 'Beyond good and evil', in *Complete Works of Friedrich Nieztsche*, vol. 5, Edinburgh: T. N. Foulis.

Nietzsche, F. (1920) *Thus Spoke Zarathustra*, New York: Random House.

Rawls, J. (1971) *A Theory of Justice*, Cambridge, MA: Harvard University Press.

Rousseau, J. J. (1762) *Du Contrat Social*, Amsterdam: Marc Michel Rey.

Rulli, T. (2019) 'Reproductive CRISPR does not cure disease', *Bioethics*, 33(9): 1071–1082.

Stone, D. (2002) *Breeding Superman: Nietzsche, Race and Eugenics in Edwardian and Interwar Britain*, Liverpool: Liverpool University Press.

Turner, B. (2006) *Vulnerability and Human Rights*, University Park, PA: Pennsylvania State University Press.

Waldron, J. (2017) *One Another's Equal. The Basis of Human Equality*, Cambridge, MA: Harvard University Press.

3

Equality and the History of Eugenics

David A. Jones

Ancient antecedents and the relation of justice to equality

The term 'eugenics' originates in the late 19th century and, as a political movement and a scientific (or pseudo-scientific) theory, it dates back no further than this. Nevertheless, some of the key elements of this modern ideology have antecedents in much earlier thought. The obvious fact of the inheritance of characteristics from parent to child had already led some ancient thinkers to propose that the health of the people could be improved by controlling which persons could marry and by the killing or exposure of newborn children deemed unfit to live.

The prime ancient example of a political system that aimed to maintain or improve the health of the population in this way was Sparta, or at least the account of Sparta given by the Greek philosopher Plutarch (c. AD 46–120s) and accepted by later Greek writers. Plutarch claimed that in Sparta all newborn infants were bathed in wine as a test of their strength and then examined by the elders to assess whether they were worthy of being reared (Plutarch *Life of Lykourgos* 16.1–3). If the child was weak, sickly, or deformed it was exposed in a ravine called the *Apothetae*. The practice of infanticide by the exposure of unwanted infants (especially of girls and of children with disabilities) was commonplace in the ancient world (Jones, 2004, pp 33ff) and Sparta was a city state renowned for its military prowess. It is thus imaginable that they placed a high value on the rearing of physically strong children who would make the best warriors.

However, the claims that *all* infants were *systematically* subject to certain tests for physical fitness, and that a determination of life or death was made by the elders, even constraining the right of the father over the child, and that all those infants who were killed were thrown into a particular ravine, are without corroboration from contemporary sources or from archaeology. There are good reasons to be sceptical of the historicity of Plutarch's account (Huys, 1996). Nevertheless, if not true it was a good story,[1] and Plutarch's Sparta provides an archetype for the ancient world of an imaginary antecedent of modern eugenics.

If Plutarch's Sparta is the imaginary exemplar, the theoretical account that comes closest to Galtonian eugenics is one set out by Plato, who considered that the ruler of the ideal Republic would seek to ensure that only the healthiest children were born. He gives the analogy of selective breeding of dogs, birds, and horses, where the best exemplars are used for breeding, and states that 'the principle holds also for mankind' (*Republic* V. 459b, translation by Shorey, 1937). In order to ensure that stocks remain healthy and do not degenerate, it is necessary to breed the 'best with the best'. The analogy with selective breeding is, of course, the foundation of Charles Darwin's account of the mechanism of evolution (Darwin, 1859, pp 20ff) and thus, indirectly, the foundation of eugenics.

Note that Plato was concerned not only with preventing procreation by the weak (negative eugenics) but also of promoting procreation among the strong (positive eugenics). Plato realized that marriage was regarded as a sacred institution, and that there are good reasons to maintain this. He also realized that people may resent being told who to marry by the ruler, especially if they were directed to marry someone inferior (however this was judged). To avoid such resentment and to maintain the sacredness of marriage, Plato imagined a religious marriage festival in which spouses would ostensibly be chosen by fate, through a fake lottery system, which would, in fact, be controlled by the ruler. This deception could be used by the ruler in a similar manner to the way that medicine is used by a physician, as a justifiable means to an end (*Republic* V. 459c–459e).

In addition to the imagined marriage lottery, Plato laid down rules for the ages of men and women to marry, so that they would be in good health when they produced young. He also stipulated that the ruler should decide how many marriages to permit so as to control the population size of the city. Children born to healthier couples would then be immediately taken from their parents to be raised by the state. Anyone who conceived a child outside these limits, that is, outside the prescribed age limits or by a union not chosen by the ruler, would be 'imposing on the state a base-born, uncertified, and unhallowed child' (*Republic* V 461b). In such a case, Plato thought that the parents should, 'preferably not even to bring to light anything whatever thus conceived', that is, procure an abortion, or if they were unable to prevent a birth should, 'dispose of it on the understanding that we cannot rear such an offspring' (*Republic* V 461c), that is, expose the infant.

Aristotle agreed with Plato on the importance of ensuring the health of the child by having age requirements for parents, and also in the city controlling its population by limiting the number of children each couple was allowed. In addition, Aristotle's view in relation to infanticide was that 'no deformed child shall be reared' (*Politics* VII.xiv, 1335b translation by Rackham, 1932) but in relation to controlling the number of children, this should not be done by infanticide 'if the regular customs hinder any of those born being exposed'. Instead, there must be a limit to the number of children and if more are conceived then 'abortion must be practised on it before it has developed sensation and life; for the line between lawful and unlawful abortion will be marked by the fact of having sensation and being alive'.

On a number of points Aristotle qualified what Plato said, in the light of his experience of what is or might be done, and what would be unrealistic or even counterproductive. Both Plato and Aristotle accepted infanticide, but Aristotle knew that it was not accepted by regular custom in some places, and so recommended early abortion as an alternative. Aristotle reserved infanticide to children who were disabled but, in this case, he thought that exposure of the infant should not only be an option but should be required by law.

In contrast to contemporary philosophers such as John Harries in the UK (in Keown, 1995, pp 8–9) or the Australian Peter Singer (1979, p 97), Aristotle held that human beings were rational by nature. A famous definition of 'person', by Boethius, an Aristotelian of the 6th century was 'an individual substance of a rational nature' (*On the two natures of Christ*, see Thomas Aquinas *Summa Theologiae* Ia 29.1) and this would apply to young or old and to people with or without disabilities.

For Aristotle, there was a fundamental inequality in how human beings were to be treated, not based on a difference in nature but due to the character of justice. According to Aristotle, justice consisted in treating equals equally and those that are unequal unequally. He sought to explore this through a distinction between 'arithmetic' equality and 'geometric' or proportional equality (*Nicomachean Ethics* V.iv, 1332a). The first gives like-for-like equivalence irrespective of the status of the persons. Thus, a fair price would be determined by the value of the item not the person buying it and the law requiring trials to be fair would apply equally to all trials. In contrast, proportional justice gives someone their due in relation to the status of that person. Thus, when distributing resources, one person might have more need than another, or one person might merit more on account of their contribution or to their place in society.

For Aristotle, proportional justice was more fundamental than arithmetic. This is because the dignity of human beings was, in the primary and more fundamental sense, unequal. Dignity, or human worth, was related to virtue and this was not possessed equally by all. Status in society did not neatly map on to virtue, for while a role might require certain virtues, there would be a dignity that applied to the role even if the person exercising this role did not possess these virtues. Nevertheless, on the whole, justice would deliver unequal results because of the inequality of need, merit, and role. From this perspective, justice applied equally to everyone only in a more generic or secondary sense. More fundamentally, people were unequal in dignity and thus should be treated by a 'proportional' equality.[2]

The conception of fundamental human equality is found in Abrahamic religions as well as in Stoicism and some other schools of philosophy, but it is not found in Aristotle. This is why Aristotle could defend a law under which disabled children were required to be killed and could defend the view that barbarians were slaves by nature. Aristotle did not deny that justice was concerned with equality, or that there were no respects in which justice applied equally to all human beings. However, he believed that, the sense in which human beings were unequal, unequal in dignity, unequal in worth or value, was more fundamental than the way in which they were equal.

In this way Aristotle, the disciple of Plato, is an antecedent of modern eugenics in its relationship to human equality. The history of eugenics shows that, as a moral or political philosophy, eugenics is based not so much on denying that there are respects in which human beings are equal but on taking as primary the respects in which human beings are unequal. It starts with traits that are present unequally in human beings, and values human beings unequally as they possess or could pass on these unequal traits.

Modern eugenics and inequality as a necessary presupposition

The term 'eugenics' was coined by Francis Galton in 1883 (Galton, 1883). Galton was, famously, the half-cousin of Darwin (sharing a common grandparent) and the concept of eugenics is very much shaped by Darwinian ideas. This is evident from the full title of Darwin's magnum opus *On the Origin of Species by Means of Natural Selection, or the Preservation of Favoured Races in the Struggle for Life* (Darwin, 1859). The term 'races' in the title refers principally to different species and subspecies of animal or plant, but it carries the clear connotation of racial distinctions between human beings. Indeed, that Darwin's interest extended beyond non-human animals is evident from his later work *The Descent of Man, and Selection in Relation to Sex* (Darwin, 1871).

The mechanisms of selection, which are at work throughout the plant and animal kingdoms and which, according to Darwin, constitute the creative forces that shape different species, are also at work in human beings. As in non-human animals, and in plants, selection always presupposes that there are inequalities between individuals and between families, breeds, or strains, and that some of these inequalities are traits that are selected for and some are traits that are selected against. In human beings, the traits possessed by 'favoured races' will be products of selection. The mechanism of selection (whether natural selection, sexual selection, selective breeding of crops and of domestic animals, or eugenic selection) does not simply presuppose inequalities between individuals or families but makes use of these inequalities as the very basis for selection. It requires, and thus *values,* the presence of inequalities.

Darwin had already applied his ideas to human beings, and others were to apply them in the political sphere: 'social Darwinists' (Rogers, 1972). Where Galton was innovative was in seeking to quantify the inherently human qualities of reasoning or intelligence, by psychometric testing, with the further aim of improving these human qualities by means of eugenic selection. This concern with measurement gave eugenics the appearance of a scientific endeavour. Eugenics did not presuppose that the improvement of traits that are heritable was sufficient to solve all the problems of society, but it did presuppose that such traits could be improved by selection (or deselection) and that this would give the human race a greater capacity to solve its problems and would add to overall human happiness.

Like Darwin, Galton was here drawing on other ideas, most notably the 'principle of population' espoused by the English clergyman Thomas Malthus (Malthus, 1798). This is implicit in the phrase 'the struggle for life' in the long title of the *Origin of Species.* Galton also drew on the idea of utility developed by the English philosopher Jeremy Bentham (for example, Bentham, 1789) which helped shape the idea that human wellbeing could be quantified through a calculus, and thus that the 'eu' in eugenics could be measured mathematically. Another part of this cultural mix was an appreciation of the striking advances in science and technology in the 19th century, the power of industrialization and, sometimes conflated, the expansion of British and European colonialism. Together these cultural factors created a context where a supposed 'science of human heredity' could be presented as a means to human progress.

Inequality and eugenic religion

Another important element in the reception of eugenics was the emergence, in the late 19th century, of the myth of a conflict between 'science' and 'religion', which allowed new ideas to be entertained even if they seemed to be in conflict with received moral or religious norms. In particular, it helped Darwinians to dispute the supposed unique dignity or status of human beings among living things, and helped eugenicists to dispute the supposed equal possession of this dignity or status by all human beings. If the idea of human equality is framed in religious terms (for example, by reference to the possession of a spiritual soul, or to being created in the image of God, or to a common supernatural end) and if inequality is presented as a presupposition or a delivery of modern science, then inequality will be regarded as the more fundamental or better-grounded principle.

It is, of course, the case that the idea of human equality can be defended on the basis of philosophy, as well as, as a religious idea. Nevertheless, the idea of equality, as it came to prominence in the 18th century, and was received in the 19th century, was typically framed in religious language.[3] This is evident, also, in the United States' Declaration of Independence:

> We hold these truths to be self-evident, that all men are created equal, that they are endowed by their Creator with certain unalienable Rights, that among these are Life, Liberty and the pursuit of Happiness.

The reference to human beings being 'created equal' and being endowed with rights 'by their Creator' are not easily dispensable from the Declaration, as the endowment is conceived of as the source of a special dignity that grounds the claims of rights. Without the appeal to some inherent dignity shared by all human beings, which is supplied in the Declaration by the reference to the Creator, it is not clear why and when and how human beings are equally worthy of respect when they are so evidently unequal in their actions and abilities.

In the late 19th and early 20th century the conflict thesis between 'science' and 'religion' took aim, primarily, at conservative forms of religion and sought to promote forms of religion that were accommodated to the scientific theories of the day and the 'new morality' that these seemed to imply.

A perfect exemplar of this accommodation is the clergyman and enthusiastic eugenicist William Inge, who was Professor of Divinity at Cambridge before being appointed Dean of St Paul's Cathedral and was one of the founders of the Eugenics Education Society in 1907. Inge argued that while eugenics, 'is part of secular morality' and has 'no supernatural sanction' it is nevertheless compatible with religion. This is so, he argued, even for Roman Catholic religion, but much more so for Protestantism. For Inge argued that, 'In Protestant countries the yoke of authority is light; and there would be no insuperable difficulties in incorporating our duty to posterity among the principles of social morality' (Inge, 1921, p 258). Eugenics could in this way be construed as a religious duty 'to posterity' (for the response of religion to more generally eugenics see Hale, 2011).

For Inge, it was thus not religion *per se* but 'primitive' 'anti-scientific' attitudes combined with 'humanitarianism' or 'sentimentalism' that led people to pursue such anti-eugenic policies as support for 'feebleminded schools, hospitals for incurables, reformatories for the morally degenerate, and prisons for the hereditary criminal' (Inge, 1921, p 265). In contrast, a religion that is open to a new morality can embrace the science of eugenics. Eugenics divides humanity into 'inferior type[s]' and 'higher types' (Inge, 1909, p 29). Inge identifies the former largely with the working classes and the latter with the upper classes. While he acknowledges that 'it would be a great mistake to suppose that *all* this [working] class belong to poor stocks whose increase is undesirable' (Inge, 1909, pp 29–30, emphasis added) he immediately qualifies this by saying that, 'on the whole, it cannot be doubted that they are inferior to the upper class, and if so, the distribution of the birth-rate in this country is highly anti-eugenic' (Inge, 1909, p 39).

Inge refers to his 'own class', he was a former assistant master at Eton, as containing 'our best families' (Inge, 1921, p 265), and opines that it is:

> [T]he duty of the well-born, in the Galtonian sense, not to cut off their own families, however dreary the outlook for their children may be. The temptation to do so will be severe; and I have no doubt that in the professional classes especially we shall have thousands of childless and servantless households, in which the tradition of culture and refined living will be maintained at the heavy price of family suicide. (Inge, 1921, p 265)

The reference to families being 'servantless' seems to reflect the idea that members of the upper classes could not envisage having children without servants to bring them up. Being unable to afford servants is equated with being unable to afford children. The war and the economic crash, together, had either killed off the 'best families' or discouraged them from breeding. A similar lament is given by Bishop Ernest W. Barnes, Anglican bishop of Oxford, writing at the end of the Second World War:

> The worst evil that can befall a nation is the destruction of its best stocks, because families belonging to these stocks hand on, generation after generation, health character and ability… The riff-raff of the community, the dull, dirty or feebleminded are useless in the army. In war-time they are overpaid in safe jobs and their numbers increase disastrously. Let the whole process go on for several generations and the whole quality of a nation declines… Total war gives a nation a 'scrub' population. (Barnes, 1945 quoted in Merricks, 2012, p 365)

To 'put away' non-persons

The same use of extremely derogatory language to describe people who are poor or who live with mental or physical disabilities, would be repeated in the postwar period by Joseph Fletcher in America. It was Fletcher, while still writing as an Episcopalian theologian and cleric, who proposed that the term 'person' be reserved for human beings who pass a threshold limit of intellectual ability. This move was overtly eugenic and was the justification for infanticide of disabled children. Writing to the father of a child with Down's syndrome, with apparent pastoral concern, Fletcher suggested that:

> People ... have no reason to feel guilty about putting a Down's syndrome baby away, whether it's 'put away' in the sense of hidden in a sanitarium or in a more responsible lethal sense. It is sad; yes. Dreadful. But it carries no guilt. True guilt arises only from an offense against a person, and a Down's [sic] is not a person. (Fletcher, 1968 cited in Jones, 2016, p 336)

This concept of personhood, which would later be popularized by Peter Singer (1979, p 97), seeks to divide human beings into persons and non-persons. The roots of this concept are clearly eugenic (as the example suggests) but bioethical discourse about 'personhood' would later be invoked in arguments about abortion (in relation to the status of the unborn child, whether disabled or healthy) and in arguments about 'brain death' and the 'vegetative state'.

One can trace the origin of modern bioethical conceptions of personhood back from Singer to Fletcher and thus back to the eugenics movement. Nevertheless, the new use of the term 'person' could sometimes be combined with a more limited form of egalitarianism – while not all human beings are equal, all 'persons' are equal. This seems to be the view of the UK bioethicist, John Harris, for example. In contrast, the eugenic conception of racial improvement has, as its starting point, the view that human beings are radically and essentially unequal, and this applies to people of above average intelligence just as much as to those of below average intelligence. For eugenics 'persons', however defined, are always unequal.

To be sure, eugenics has often been driven by a visceral fear of people with more profound disabilities or who are perceived as socially disruptive, as evident in the list of undesirables given by Inge or Barnes. However, these negative traits are not equally bad and not present equally in individuals, just as purported positive traits are not equally good or present equally in individuals. Everything is present by degrees. Despite Inge's simple division of people of families into the 'inferior' and the 'higher', the essential character of eugenics is not to value persons at all but rather to value traits, and this valuing will seldom if ever be 'equal' between persons.

Eugenics and the democratic spirit

The Welsh academic lawyer and eugenicist Glanville Williams, writing in 1957, defended the legalization of sterilization, abortion, infanticide, and euthanasia as means of 'racial improvement'. In this context, he quoted with approval the statement of Betrand Russell that '…the ideas of eugenics are based on the assumption that men are unequal, while democracy is based on the assumption that they are equal' (Williams, 1957, p 73 quoting Russell, 1929, p 262). However, this statement was not put forward as a criticism of eugenics but as a criticism of democracy. 'It is, therefore, politically very difficult to carry out eugenic ideas in a democratic community when those ideas take the form, not of suggesting that there is a minority of inferior people such as imbeciles, but of admitting [sic] that there is a minority of superior people' (Williams, 1957, p 73).

The eugenics movement, from its origins, thus faced opposition or resistance from other modern political developments which were egalitarian in spirit, most notably by the expansion of democracy. The expansion of democracy, both in relation to the number of countries with democratic systems of government and in relation to the expansion of the franchise, embodies a valuing of human beings that is radically at odds with that of eugenics. Without denying diversity, and thus all manner of inequalities, democracy begins with valuing what is common.

This radical difference in underlying philosophy is well-expressed by the English writer, G.K. Chesterton, whose political ideas mixed conservatism and radicalism, somewhat in the tradition of his fellow writers and compatriots Samuel Johnson and William Cobbett, both of whom he admired. Chesterton was one of the most outspoken critics of eugenics which he described as 'a war upon the weak' (Chesterton, 1922, p 64). This attitude echoed the way, a century earlier, Cobbett had railed against Malthusian attempts to prevent the poor working classes from marrying while proposing no such restrictions for the idle upper classes. To understand Chesterton's anti-eugenic works it is best to start with earlier writings where he sets out his positive view of life and his defence of democracy as a system of government (note, this was before the universal franchise).

> [T]he things common to all men are more important than the things peculiar to any men. Ordinary things are more valuable than extraordinary things; nay, they are more extraordinary. Man is something more awful than men; something more strange… Death is more tragic even than death by starvation. Having a nose is more comic even than having a Norman nose.
>
> This is the first principle of democracy: that the essential things in men are the things they hold in common, not the things they hold separately. And the second principle is merely this: that the political instinct or desire is one of these things which they hold in common. [It is not like]… being Astronomer Royal, and so on. For these things we do not wish a man to do at all unless he does them well. It is, on

> the contrary, a thing analogous to writing one's own love-letters or blowing one's own nose. These things we want a man to do for himself, even if he does them badly. (Chesterton, 1908, p 81)

This concept that what human beings 'hold in common' is of the greatest value is precisely what is lacking in eugenic thinkers for whom it is what *distinguishes* people from one another that is the basis of value, and is either selected for or selected against.

It might also be noted that while eugenicists have sometimes appealed to utilitarian ideas, or to economic ideas associated with free market thinkers such as the Scottish economist Adam Smith, each of these ideas, in their way, also embody the concepts of equality that are in tension with the thought-world of eugenics.

Bentham, for example, was a kind of egalitarian, not only in that every one counted as one and none as more than one, but also in that pleasures and pains were not distinguished between 'higher' and 'lower':

> Prejudice apart, the game of push-pin is of equal value with the arts and sciences of music and poetry. If the game of push-pin furnish more pleasure, it is more valuable than either. Everybody can play at push-pin: poetry and music are relished only by a few. The game of push-pin is always innocent: it were well could the same be always asserted of poetry. (Bentham, 1825, p 208)

The humour in these remarks (for example, in relation to the non-innocence of poetry) and the appeal to common pleasures could easily have been written by Chesterton. This is the democratizing of pleasure with no automatic preference for the 'tradition of culture and refined living' that Inge (1921, p 265) wished to preserve for the few through use of more or less coercive eugenic measures against the many.

Equality is also fundamental to the economic vision of Adam Smith, not so much equality of pleasures or equality of outcome as equality of opportunity (Peart and Levy, 2005). Thus, the emphasis on social and economic factors in shaping society led Smith to downplay the significance of biological inheritance, at least in general and for the most part:

> The difference of natural talents in different men is, in reality, much less than we are aware of; and the very different genius which appears to distinguish men of different professions, when grown up to maturity, is not upon many occasions so much the cause, as the effect of the division of labour. The difference between the most dissimilar characters, between a philosopher and a common street porter, for example, seems to arise not so much from nature, as from habit, custom, and education.

> The difference... widens by degrees, till at last the vanity of the philosopher is willing to acknowledge scarce any resemblance. (Smith, 1776, I.2)

The emphasis on social determinants of inequality is also a feature of thinkers on the left of politics, from Marxists to communitarians (evident in the critique of eugenics offered by Richard Lewontin (1991), Stephen Jay Gould (1996), or Hilary and Steven Rose (2000)), notwithstanding notable leftwing supporters of some forms of eugenics such as the geneticist, J.B.S. Haldane (in his earlier life) and the Irish playwright George Bernard Shaw (Paul, 1984). Thus, economic ideas whether of the right or of the left tend to lead to caution about overstating the role of biological inheritance as an explanation for current social inequalities.

The bland assumption of Inge that the 'best families' are to be found in the upper classes and their social position is a reflection of being 'well-born, in the Galtonian sense' (Inge, 1921, p 265) would founder equally on the rock of libertarianism or the whirlpool of socialism. There were, of course, thinkers of the left and of the right who embraced eugenics, and especially in the 1920s, but the movement was inhibited, not only by religious conservatism (by Roman Catholics) but also by a countervailing spirit of democracy, human rights, and belief in equal human dignity which was common to right and left.

Contrasting equalities contrasting inequalities

It might be thought that the United States, founded on a declaration of fundamental equality, suspicious of government intervention in private life, attracted to the ideal of equality of opportunity, and without a hereditary monarchy or aristocracy, would be better able than the 'old world' to resist the eugenic ideas of Galton. However, this is far from the case.

Eugenic ideas proved highly popular in the United States (Lombardo, 2011; Reilly, 2015). Already in 1896, the State of Connecticut had passed a law to prohibit marriage by anyone who was 'epileptic, imbecile or feeble-minded'. In 1907 Indiana passed a more draconian law which would involve the compulsory sterilzation of 'confirmed criminals, idiots, imbeciles and rapists'. This law was used as the model by other states, and eventually 32 States enacted similar legislation. The practice of compulsory sterilization of the unfit would continue into 21st century with tens of thousands of Americans being sterilized without their consent.

One concern that seems to have encouraged people to embrace eugenic ideas was anxiety on the part of Americans of North European descent about immigration and about the increase in population of other ethnic or racial groups. While the American Constitution appealed to a fundamental human equality, it was written primarily by Northern European Protestants (and Deists) and some from this constituency expressed a concern that the political system of the United States would be under threat if America absorbed too many Roman Catholics and/or people from (other) 'primitive' cultures. Linked with this concern was the idea that American civilization presupposed a particular biological inheritance at the level of race or ethnicity, and that this would be undermined by intermarriage with 'inferior' races. As early as the 17th century, some States had prohibited marriage between people of European ancestry and African Americans or Native Americans, and by the 19th century most States had 'anti-miscegenation' laws.

While the civil war overturned the institution of slavery it did not lead to an immediate overturning of laws against interracial marriage, which had not been confined to the Confederate States. Such laws remained in force in most States in the United States until after the Second World War. Indeed, they increased in the 20th century. It was only with the civil rights movement in the 1960s that there was a concerted attempt to overturn them across the United States.

Racial thinking increased internationally in the late 19th century. In America, 'Jim Crow' laws reflected increasing recourse to racial explanations for social problems. In the late 19th and early 20th century, eugenic ideologies provided people with a respectable scientific rationale for racism in general and for anti-miscegenation laws in particular. This may explain why the United States turned to sterilization and the restriction of marriage as a means to prevent the birth of people with disabilities or with socially undesirable traits (rather than turning to institutionalization as in England).

Despite the very prominent affirmation of equality in the founding documents of the United States, the efforts to entrench social practices of racial inequality, including by laws restricting marriage, provided a cultural environment within which eugenic ideas were welcomed. It was also the case that the eugenic laws, while framed by reference to people with intellectual disability or recurrent anti-social behaviour, and not overtly by reference to racial categories *per se*, were not applied equally to different racial and ethnic groups. They were used disproportionately to restrict the procreative freedom of African American, Hispanic, and Native American women (Bocquillon, 2018; Reilly, 2015; Lombardo, 2011).

The eugenics movement in America also benefited from support from rich philanthropists such as Andrew Carnegie, John D. Rockefeller, and John Harvey Kellogg. These helped support research centres such as the Eugenics Record Office, founded in 1910 by the American zoologist Charles Davenport, and the Race Betterment Foundation, founded in 1914 by Kellogg. Eugenics was well suited to funding by wealthy patrons, who were inclined to see themselves as 'well-born in the Galtonian sense', whether or not they had inherited wealth.

Another feature in the expansion of eugenics across the United States was the birth control movement. This movement, associated with figures such as the American writer Margaret Sanger, was tinged with fear of population increase among the poor and especially among non-white Americans. It was overtly eugenic. Nevertheless, insofar as it was also framed as improving the lives and status of women, eugenic birth control could be presented as enhancing equality even as it was reinforcing racial inequality.

While eugenics might have been expected to face strong headwinds in the United States, it had certain obvious advantages in England. The principle of heredity was embodied in the heredity monarchy and in the House of Lords. Both Darwin and Galton were English and, by the turn of the century, both were celebrated for their intellectual achievements. Rather than a civil war in the mid-19th century, which at least raised the issue of the unequal treatment of African Americans as slaves, the expansion of British influence and wealth through its empire in the 19th century was popular and uncontroversial in Britain. Britain also had a birth control movement, led by Marie Stopes, who like Margaret Sanger was a keen advocate of eugenics.

At the beginning of the 20th century, eugenics enjoyed support from across the political spectrum in the United Kingdom both on the right and on the left (Paul, 1984). In 1904, the Conservative Government established a Royal Commission on the Care and Control of the Feeble-Minded. In 1907, the Eugenics Education Society was founded and in 1912 the First International Eugenics Congress was held in London. The President of the Congress was Major Leonard Darwin, son of Charles Darwin, while Winston Churchill was a Vice-President. Other attendees included Alexander Graham Bell, the celebrated physician Sir William Osler and the former Prime Minister Arthur Balfour (BMJ, 1912).

The Royal Commission report was published in 1908, and in 1912 a private member's bill was introduced to parliament to implement the proposals: the Feeble-Minded Control Bill. This was influenced by the Indiana law of 1907. It would have enforced compulsory sterilization of those deemed 'feeble-minded' and would have taken away their right to marry. However, thanks in part to public criticism by Chesterton, the 'feeble-minded bill' was withdrawn. In its place, the Liberal government brought forward the Mental Deficiency Bill 1913. Churchill, who had crossed the parliamentary floor in 1904 and was at this point a Liberal, had a hand in drafting the government bill. It did not include sterilization, which Churchill himself favoured, but it allowed the indefinite incarceration of 'idiots, imbeciles, the feeble-minded and moral defectives'.

The bill was presented as a compromise and attracted strong cross-bench support. In the final vote it was opposed by only three MPs. Nevertheless, one of these, Josiah Wedgwood was unrelenting in his opposition and attempted a filibuster, albeit without ultimate success. Wedgwood attacked the bill as an unjust deprivation of liberty but his disillusion with his party (the Liberal Party) began earlier with its failure to address land reform or female suffrage. Like Chesterton, his opposition to these eugenic proposals had its roots in a certain democratic spirit. While Chesterton and Wedgwood represented minority voices in the public and in Parliamentary debates, their opposition drew public attention to the way that eugenic proposals were in tension with democratic conceptions of equality. They also may have muted its worst excesses.

Eugenic laws were thus passed in the United States and in Britain but they advanced further in the United States than they did in Britain, which ultimately did not embrace compulsory sterilization. In both countries the process of acceptance of eugenics can be understood in relation to the principle of inequality on which eugenics relies. In America, the process of acceptance was facilitated by positive interaction with the existing inequality of racial laws and practices masked, to some extent, by the equality concerns of the birth control movement. In Britain eugenics was inhibited, at least to some extent, through adverse interaction with advocates of equality rooted in political liberty and democracy.

A fall from favour and the rise of the new eugenics

The adoption of eugenic policies in Nazi Germany, and their ultimate expression in the mass killing of disabled children and adults in the *Aktion T4* programme, was influenced directly by ideas emerging in the UK and in America. Adolf Hitler quoted the example of the 1907 Indiana law in *Mein Kampf* and the German medical establishment was strongly influenced by eugenic ideas long before the rise of National Socialism in the 1930s (Rubenfeld and Sulmasy, 2020). It is true that Nazi eugenics was more viscerally antisemitic than American or British forms of eugenics, but both these older forms were already overtly racist and embedded in anti-miscegenation laws and colonialism, respectively. As argued above, eugenics was founded on an appeal to inequality.

The association of eugenics with Nazism weakened the eugenic movement in the period immediately after the Second World War, and inhibited some of the rhetoric that had been common in the first half of the twentieth century (see Glover, 2006, though for exceptions note previous quotes from Barnes, 1945, Williams, 1957, and Fletcher, 1968). However, a new, less overt, eugenics movement began to emerge in the 1960s. Perhaps the most important factor in this new phase of the eugenics movement was the discovery in 1953 of the structure of DNA. This discovery paved the way for genetic testing for hereditary diseases, including prenatal testing.

The 1960s also saw a rise in the fear of global overpopulation which was focused, in particular, on the global South (Erlich, 1968). At the same time, the birth control movement received a major impetus from the development of a convenient and effective form of contraception: the contraceptive pill. This technical development was followed by a gradual shift in social attitudes to the acceptability of abortion if contraception should fail.

In Great Britain, access to abortion was widened by the *Abortion Act 1967*. This statute included an overtly eugenic indication for abortion: the presence of 'a substantial risk that if the child were born it would suffer from such physical or mental abnormalities as to be seriously handicapped' (S. 1(b)). The background to this clause was the thalidomide disaster. Thalidomide was a sedative drug launched in 1959 but was soon found to cause serious disability in children if taken during pregnancy. The drug was withdrawn, but not before hundreds of children had been affected. There was great public sympathy for women who had taken the drug and then wished to have an abortion rather than give birth to a disabled child. According to the advocates of abortion, 'thalidomide was the motor that reinvigorated the Abortion Law Reform Association and which paved the way for reform' (Hindell and Simms, 1971, p 108 cited in Jones, 2004, p 204).

The possibilities for genetic control of reproduction were further expanded in 1989 with the development of pre-implantation genetic diagnosis (PGD) for screening of children conceived by *in vitro* fertilization. It is clear that the possibility of preventing the birth of disabled children played an important role both in the passage of the *Abortion Act 1967* and in the *Human Fertilisation and Embryology Act 1990*.[4] During the debates that led up to the vote on that latter legislation, Progress Campaign, a group campaigning against restrictions on human embryo research, arranged for 200 families affected by genetic disease to visit Parliament. As a result of such activities, 'in the crucial debates in late 1989 and early 1990, 75 per cent of those arguing for embryo research made significant reference to its potential contribution to the prevention of genetic disorder' (Mulkay, 1997, p 63).

The new eugenics[5] sought to distinguish itself from the earlier history of eugenics by framing eugenic choices as a matter of parental choice or of women's reproductive choice, rather than a matter of coercive state policy. However, while this emphasis clearly differentiates the new eugenics from Nazi policies, it is harder to differentiate it from the origins of British and American eugenics. Not all eugenicists in this earlier period favoured coercive means. Their first aim was to encourage support for eugenic ideas both in private decisions and in public policy and most were pragmatic about how to increase eugenic outcomes across society.

It is also disingenuous to claim that prenatal screening and PGD are simply about increasing reproductive choice as though the offering of choice was somehow neutral between outcomes. The great societal investment of time, effort, and resources, both scientifically and politically, in making certain genetic choices available is because eugenic choices were thought to be beneficial, both at an individual level and at a social as well as economic level. If preventing the birth of a child with an inherited disability were not thought beneficial then the test would not be developed or offered. It may be said that efforts to promote both suppression of fertility (through contraception) and promotion of fertility (by assisted reproduction) show that medical technology does not aim to favour one outcome but aims to increase choice. However, it is evident that the historic drive to develop contraception was heavily influenced by eugenic motivations (aiming at avoiding undesirable births) as was the development of assisted reproduction (aiming to promote the birth only of healthy children). It is clear that the outcomes are not viewed equally, as it is only the presence of a gene associated with disease that is grounds for destroying the foetus or embryo, not the absence of such a gene. It is the presence of a disability that is specific grounds for abortion, not the absence of a disability. It is this fundamental, and ultimately lethal, inequality that marks out prenatal testing and PGD as eugenic practices.

The key moral or political question is whether what human beings hold equally in common, the fact of a shared humanity, has priority over the way in which one human being can excel another in physical or intellectual strength and ability. What is common has priority within the Abrahamic tradition, in some schools of ancient philosophy, and in conceptions of democracy and equal human rights. The contrary view is held in various forms of aristocracy, elitism, pseudo-scientific racism, and ablism, of which eugenic ideologies, both old and new, are but recent variants. Rejection of human equality is the very essence of eugenics.

Notes

1 '*Se non è vero, è molto ben trovato*', a saying popularized by the Italian esoteric thinker Giordano Bruno.

2 This insight into the importance of Aristotle's conception of proportional equality for explaining his acceptance of injustices such as infanticide and 'natural' slavery was pointed out to me by Kevin Flannery (see Jones, 2009, p 14, n. 15).

3 For example, the philosophy of Immanual Kant, which has had a strong influence on modern ideas of equality and dignity, also has clear echoes of religious language, even if it seeks to be a form of religion 'within the bounds of bare reason' (Kant, 1793).

4 For example, Enoch Powell MP, who had sponsored a Bill in 1985 to ban human embryo research was nevertheless in favour of selection of embryos on eugenic grounds (see Theodosiou and Johnson, 2011 cited in Jones, 2017, p 90), and this fractured the opposition to the Human Fertilisation and Embryology Bill.

5 For further discussion of the 'new eugenics' see MacKellar and Bechtel (2014) and Chapters 13 and 14 in the present volume.

References

Aquinas, T. (1911) *Summa Theologica*, trans. Fathers of the English Dominican Province, London: Burns, Oates (reprint 1981, New York: Thomas More Publishing).

Barnes, E.W. (1945) 'University of Cambridge Sermon', in *The Papers of E.W. Barnes*, 22 April 1945, 12/1/649.

Bentham, J. (1789) *An Introduction to the Principles of Morals and Legislation*, London: T. Payne and son.

Bentham, J. (1825) *The Rationale of Reward*, London: John and H. L. Hunt, trans. R. Smith from J. Bentham (1811) *Théorie des Peines et des Récompenses*, London: De l'imprimerie de Vogel et Schulze.

BMJ (1912) 'First International Eugenics Congress', *British Medical Journal*, 2 (2692): 253–255.

Bocquillon, S.P. (2018) 'Sterilization in the United States: The dark side of contraception', *Revue de recherche en civilisation américaine* [*En ligne*], 8, 2018, 17 December, Available from: http://journals.openedition.org/rrca/1169 [Accessed 25 January 2025].

Chesterton, G.K. (1908) *Orthodoxy*, London: John Lane.

Chesterton, G.K. (1922) *Eugenics and Other Evils*, London: Cassell and Co.

Darwin, C.R. (1859) *On the Origin of Species by Means of Natural Selection*, London: John Murray.

Darwin, C.R. (1871) *The Descent of Man, and Selection in Relation to Sex*, London: John Murray.

Ehrlich, P.R. (1968) *The Population Bomb*. New York: Ballantine Books.

Fletcher, J. (1968) 'The Right to Die', *Atlantic Monthly* (April): 59–64.

Galton, F. (1883) *Inquiries into Human Faculty and Its Development*, London: Macmillan.

Glover, J. (2006) 'Nazi Eugenics, Virginia Woolf and the Morality of Designer Babies', *The Guardian*, 6 May.

Gould, S.J. (1996) *The Mismeasure of Man: Revised Edition*, New York: W.W. Norton.

Hale, F. (2011) 'Debating the new religion of eugenics: Catholic and Anglican positions in early twentieth-century Britain', *Heythrop Journal*, 52(3): 445–457.

Hindell, K. and Simms, M. (1971) *Abortion Law Reformed*, London: Peter Owen.

Huys, M. (1996) 'The Spartan practice of selective infanticide and its parallels in ancient Utopian tradition' *Ancient Society*, 27(1996): 47–74.

Inge, W.R. (1909) 'Some moral aspects of eugenics', *Eugenics Review*, 1(1): 26–36.

Inge, W.R. (1921) 'Eugenics and religion', *Eugenics Review*, 12(4): 257–265.

Jones, D.A. (2004) *The Soul of the Embryo: An Enquiry into the Status of the Human Embryo in the Christian Tradition*, London: Continuum.

Jones, D.A. (2009) 'Incapacity and personhood: Respecting the non-autonomous self', in H. Watt (ed), *Incapacity and Care: Controversies in Healthcare and Research*, Oxford: The Linacre Centre, pp 8–23.

Jones, D.A. (2016) 'Apostles of suicide: Theological precedent for Christian support of "assisted dying"', *Studies in Christian Ethics*, 29(3): 331–338.

Jones, D.A. (2017) 'Editing out the embryo: the debate over "human genome editing" in the United Kingdom and the United States', *National Catholic Bioethics Quarterly*, 17(1): 83–105.

Kant, I. (1793) *Die Religion innerhalb der Grenzen der bloßen Vernunft*, Königsberg: den Friedrich Nicolovius, English translation by Pluhar, W.S. *Religion within the Bounds of Bare Reason.* Indianapolis: Hackett Publishing Company, 2009.

Keown, J. (1995) *Euthanasia Examined*, Cambridge: Cambridge University Press.

Lewontin, R.C. (1991) *Biology as Ideology: The Doctrine of DNA*, New York, NY: Harper Perennial.

Lombardo, P.A. (ed) (2011) *A Century of Eugenics in America: From the Indiana Experiment to the Human Genome Era*, Bloomington, IN: Indiana University Press.

MacKellar, C. and Bechtel, C. (eds) (2014) *The Ethics of the New Eugenics*, New York: Berghahn Books.

Malthus, T. (1798) *An Essay on the Principle of Population*, London: J. Johnson.

Merricks, P.T. (2012) '"God and the Gene": E. W. Barnes on eugenics and religion', *Politics, Religion and Ideology*, 13(3): 353–374.

Mulkay, M. (1997) *The Embryo Research Debate: Science and the Politics of Reproduction*, Cambridge: Cambridge University Press.

Paul, D. (1984) 'Eugenics and the left', *Journal of the History of Ideas*, 45(4): 567–590.

Peart, S. and Levy, D.M. (2005) *The "Vanity of the Philosopher": From Equality to Hierarchy in Post-Classical Economics*, Ann Arbor, MI: University of Michigan Press.

Rackham, H. (1932) *Aristotle Politics: Loeb Classical Library*, Cambridge, MA: Harvard University Press.

Reilly, P.R. (2015) 'Eugenics and involuntary sterilization: 1907–2015', *Annual Review of Genomics and Human Genetics*, 16(1): 351–368.

Rogers, J.A. (1972) 'Darwinism and social Darwinism', *Journal of the History of Ideas*, 33(2): 265–280.

Rose, H. and Rose, S. (eds) (2000) *Alas, Poor Darwin: Arguments Against Evolutionary Psychology*, London: Johnathan Cape.

Rubenfeld, S. and Sumasy, D.P. (eds) (2020) *Physician-Assisted Suicide and Euthanasia Before, During, and After the Holocaust*, Lanham, MD: Lexington Books.

Russell, B. (1929) *Marriage and Morals*, New York: Liveright Publishing.

Shorey, P. (1937) *Palto Republic I: Loeb Classical Library*, Cambridge, MA: Harvard University Press.

Singer, P. (1979) *Practical Ethics*, Cambridge: Cambridge University Press.

Smith, A. (1776) *An Inquiry into the Nature and Causes of the Wealth of Nations*, London: W. Strahan and T. Cadell.

Theodosiou, A.A. and Johnson, M.H. (2011) 'The politics of human embryo research and the motivation to achieve PGD', *Reproductive Biomedicine Online* 22(5): 457–471.

Williams, G.L. (1957) *The Sanctity of Life and the Criminal Law*, New York: Alfred A. Knopf.

4

Changing Understandings of Equality in the Human Rights Debate on Heritable Genome Editing

Britta van Beers

Introduction

Human rights are at the heart of the legal-ethical governance of biomedical and genetic technological developments. Already since the 1990s, human rights organizations such as the Council of Europe and UNESCO have taken the lead in creating legal instruments to address the complex issues raised by these technologies. Within these documents, several uses of biomedical technologies are declared off-limits, including not only reproductive cloning, sex selection, and organ selling, but also heritable genome editing. The main concern underlying existing human rights bans on heritable genome editing is that this technology may lead to practices that are at odds with human dignity.

However, now that CRISPR-Cas9 has brought heritable genome editing within reach, the call for a novel, more permissive human rights-based approach to heritable genome editing is growing louder, especially among members of the science community. Within the proposed frameworks, human dignity is fading into the background, while equality appears to be on the rise as normative foundation (section 'From dignity to equality').

Like human dignity, equality is a rich but complex notion, giving rise to multiple readings. In this chapter, I will examine the two main shapes that the principle of equality has, so far, taken within the human rights governance of biogenetics: equality as protection against genetic discrimination and eugenic practices, and equality as equitable access to the fruits of biogenetic science (section 'Two dimensions of equality'). Traditionally, the first interpretation, that emphasizes the anti-discriminatory and anti-eugenic aspect of equality in the context of human genetics, has dominated in human rights approaches to heritable genome editing, resulting in a prohibitive stance. However, since the birth of genetically modified twins in China in 2018, the reading of equality as an individual right to equitable access is increasingly invoked, especially by those who favour a regulatory, instead of prohibitive, approach to heritable genome editing (section 'Human germline gene editing in the name of equality').

I will argue that, although equality as equitable access is of great importance for the governance of biogenetic technologies, this emerging egalitarian approach comes with certain risks and dangers in the context of heritable genome editing and can lead to results that are at odds with equality in its first dimension (section 'Shortcomings of the equal access approach from a human rights perspective'). In order to find a better balance between both dimensions of equality in this context, it is necessary, I suggest, to reconnect equality with the human rights principle of human dignity (section 'Conclusion: 'equal in dignity'').

From dignity to equality

Since the 1990s, human rights have come to play a central role in the governance of biomedical and genetic technologies (Andorno, 2009). The most prominent international human rights instruments in the field of biomedical law are the Council of Europe's *Convention on Human Rights and Biomedicine*, also known as the Oviedo Convention, and UNESCO's *Universal Declaration on the Human Genome and Human Rights*, both from 1997. According to the Explanatory Report of the Oviedo Convention, human rights concerns about biomedical developments need to be addressed on three levels, namely (1) the individual, (2) the societal, and (3) the human species levels (Explanatory Report to the Oviedo Convention, 1997, sub 14). Consequently, even though human rights, such as the right to physical integrity and privacy, are essential for biolaw and bioethics, human rights protection in this field is about more than the protection of the individual and individual rights. Given the fact that 'progress in knowledge of the genome is producing more ways of influencing and acting on it', according to the Explanatory Report (sub 14), it is no longer only 'the individual or society that may be at risk but the human species itself'. Consequently, within the Oviedo Convention, 'reference is made to the benefits to future generations and to all humanity, while provision is made throughout the text for the necessary legal guarantees to protect the identity of the human being' (Explanatory Report to the Oviedo Convention, 1997, sub 14). More generally, the

Explanatory Report (sub 9) identifies human dignity as 'the essential value to be upheld' in the Oviedo Convention.

Likewise, UNESCO's Universal Declaration on the Human Genome and Human Rights (hereafter: UDHGHR) highlights not only individual rights and societal interests that need to be taken into consideration in the governance of biogenetic technologies, but also principles and interests related to humanity, as a whole. In this vein, according to Article 1:

> [T]he human genome underlies the fundamental unity of all members of the human family, as well as the recognition of their inherent dignity and diversity. In a symbolic sense, it is the heritage of humanity.[1]

Heritable genome editing offers one of the most striking illustrations of the far-reaching way in which 'knowledge of the genome is producing more ways of influencing and acting on it', to use the previously cited words from the Oviedo Convention's Explanatory Report. In the case of heritable genome editing, the stakes are considered to be so high, that Article 13 of the Oviedo Convention categorically prohibits this type of genetic modification.[2] The Explanatory Report (sub 89) offers the following reasons for this prohibition:

> Whilst developments in this field may lead to great benefit for humanity, misuse of these developments may endanger not only the individual but the species itself. The ultimate fear is of intentional modification of the human genome so as to produce individuals or entire groups endowed with particular characteristics and required qualities.

In a similar vein, concerns about the impact of heritable genome editing on human dignity can be recognized in the context of UNESCO's UDHGHR, even though the Declaration does not contain an explicit ban on it. Instead, Article 24 of this declaration refers to 'germ-line interventions' as one of the 'practices that could be contrary to human dignity'. According to the same provision, UNESCO's International Bioethics Committee (hereafter: IBC) has a special task in the identification of such possibly dehumanizing practices, as part of its more general responsibility to give advice concerning the follow-up of the Declaration. In 2015, the IBC issued a report, thereby 'updating its reflection on the human genome and human rights' (see title of IBC, 2015). In this report, building on the human genome as the heritage of humanity (Article 1 UDHGHR) and on human dignity, the IBC (2015, sub 107) indeed identifies heritable genome editing as contrary to human dignity:

> [...] interventions on the human genome should be admitted only for preventive, diagnostic or therapeutic reasons and *without enacting modifications for descendants*, as affirmed in Article 13 of the Oviedo

> Convention. The alternative would be to *jeopardize the inherent and therefore equal dignity of all human beings and renew eugenics*, disguised as the fulfilment of the wish for a better, improved life. (emphasis added)

In short, within the context of both these human rights documents, grave concerns at the level of humanity are expressed about the prospect of genetically modifying offspring (Andorno, 2009; Van Beers, 2020).

However, since the rise of the gene editing tool CRISPR-Cas9, and even more so since the birth of genetically modified baby girls in China, 'a novel human rights-based approach' (Kleiderman, Ravitsky, and Knoppers, 2019, p 512) to heritable gene editing is being proposed, with a different normative focus and a different conclusion on the permissibility of this reprogenetic technology. According to its proponents, what is needed from a human rights-based point of view, is not a ban on heritable genome editing, but a translational pathway towards making this technology safe for clinical use in order to eliminate serious genetic diseases.

Compared to the prohibitive approach that can be found in existing international human rights documents, the new reading of the meaning of human rights for heritable genome editing is characterized by a shift away from collectively towards more individually oriented values and interests (van Est et al, 2016; van Beers, 2020), dismissing notions such as human dignity and the heritage of humanity, and emphasizing notions such as reproductive autonomy, patients' rights and individual rights such as the human rights to health and to science. In addition, a more medical outlook comes to the fore within this novel approach, stressing medical standards such as medical safety, medical protocols, and medical 'therapeutic' purposes for which heritable genome editing could be used ('serious genetic diseases').

This approach is not only propagated by several bioethics and biolaw scholars (Alta Charo, 2018; de Miguel Beriain, 2018; Kleiderman, Ravitsky, and Knoppers, 2019; Boggio, Romano, and Almqvist, 2020; Boggio and Yotova, 2021) but can also be recognized in documents from national scientific and medical-professional bodies, academies and councils, such as the US National Academies of Science (NASEM, 2017), the UK Royal Society (NASEM/Royal Society, 2020), the European Society of Human Genetics and the European Society of Human Reproduction and Embryology (de Wert et al, 2018). In this vein, the NASEM, in its 2017 report on human genome editing, describes its own normative framework as follows:

> [F]ocused on principles that are aimed at protecting and promoting the health and well-being of individuals; approaching novel technologies with careful attention to constantly evolving information; respecting individual rights; guarding against unwanted societal effects; and

> equitably distributing information, burdens, and benefits. (NASEM, 2017, p 32)

In other words, of the three levels that the Oviedo Convention and the UDHGHR take into account (humanity, society and the individual), only the level of the individual and society are deemed relevant by the NASEM.

Another striking illustration of the 'novel human rights-based approach' that is currently emerging among national scientific and medical-professional bodies, is the 2018 report *Genome editing and human reproduction: social and ethical issues* from the UK Nuffield Council of Bioethics. Importantly, for its ethical reflection on heritable genome editing, the Council relies on the concepts and language of human rights (Nuffield Council, 2018, sub 3.2). Moreover, similarly to the normative framework of the Oviedo Convention, the report discusses heritable genome editing at the level of the individual, society, and humanity, in three separate sections. Nevertheless, in the end, the Council comes to very different recommendations with regard to heritable genome editing than the Council of Europe and UNESCO.

To start with the level of humanity, the Nuffield Council dismisses arguments related to humanity as being too 'nebulous' (Nuffield Council, 2018, sub 4.57) and as guilty of 'genomic essentialism' (Nuffield Council, 2018, sub 3.82). The Council's line of reasoning is that 'entitlement to human rights does not depend on the possession of a 'human genome', even if such a thing could be described' (Nuffield Council, 2018, sub 3.90). Therefore, it does 'not find the concept of human dignity helpful in this context' (Nuffield Council, 2018, sub 5.16). As a result, according to the Council, only considerations and principles related to the levels of the individual and society matter in this discussion (Nuffield Council, 2018, sub 3.90). Based on these, the main conclusion of the report is 'that so long as heritable genome editing interventions are consistent with the welfare of the future person and with social justice and solidarity, they do not contravene any categorical moral prohibition' (Nuffield Council, 2018, sub 5.16). In other words, a categorical ban is off the table because 'heritable genome editing interventions may be morally permissible in certain circumstances' (Nuffield Council, 2018, sub 5.17).

As to the welfare of the future person, the Council states that 'there is no a priori reason that preferences beyond the avoidance of disease should not also be consistent with the welfare of the future person' (Nuffield Council, 2018, sub 3.46). In practice, this means that the Council regards even germline interventions for the purpose of human enhancement as compatible with human rights, as long as the technology is medically safe and as long as use of the technology does not 'produce or exacerbate social division or the unmitigated marginalisation or disadvantage of groups within society' (Nuffield Council, 2018, pp 87 and 157). Since the Council assumes that the technology will become 'clinically safe in the foreseeable future' (Nuffield Council, 2018, sub 5.5), its most pressing remaining concern about heritable genome editing appears to be the effects of this technology on equality.

All in all, when it comes to the protection of more collective interests and values within the proposals for a novel human rights-based approach to heritable genome editing, equality appears to have taken over from dignity as a normative foundation. Equality, like the related notion of human dignity, is a rich, multi-dimensional legal-ethical principle that is indispensable for human rights-based approaches to heritable genome editing, bringing together concerns at the level of both society and the individual. Consequently, equality has multiple dimensions in the context of heritable genome editing. However, as I shall discuss, also at the level of equality, a shift from collectively towards more individually oriented values and interests can be detected within the recent proposals for a novel human rights-based approach.

Two dimensions of equality

Given the increasing prominence of equality in public debates about heritable genome modification, the question as to what this principle means for the human rights governance of biomedical developments is becoming more pressing. Without a doubt, the idea of equality is very closely connected to dignity. As Article 1 of the Universal Declaration of Human Rights famously states: 'All human beings are born free and *equal in dignity* and rights' (emphasis added). Yet, what equality means, exactly, in the context of biomedical technologies is much less clear.

A closer look at the Council of Europe's Oviedo Convention and UNESCO's UDHGHR suggests that, generally, equality has two important dimensions in human rights frameworks for biogenetics: (1) equality as a protection against discriminatory and eugenic practices, and (2) equality as equitable access to the benefits of biogenetic science.

As to the *first dimension*, the principle of equality can be recognized in provisions that aim to protect against genetic discrimination, genetic determinism, and eugenic practices. In this vein, Article 11 of the Oviedo Convention prohibits 'any form of discrimination against a person on grounds of his or her genetic heritage', in the same way that Article 6 of the UDHGHR proclaims that 'no one shall be subjected to discrimination based on genetic characteristics that is intended to infringe or has the effect of infringing human rights, fundamental freedoms and human dignity'.

Here, equality is understood as entailing a right to equal respect, regardless of genetic, biological, or factual differences. As such, it is clear that this dimension of equality presupposes a non-empirical view of the human that goes back to a certain normative understanding of humanity. What connects individuals according to the philosophy of human rights, based on which they deserve to be treated as equals, is their common humanity, that is, the dignity with which they are born. This counterfactual, anti-reductionist view of the human comes to powerful expression in Article 2 of the UDHGHR:

(a) Everyone has a right to respect for their dignity and for their rights regardless of their genetic characteristics.
(b) That dignity makes it imperative not to reduce individuals to their genetic characteristics and to respect their uniqueness and diversity.

Accordingly, human dignity can be regarded as a 'humanizing fiction' (Labrusse-Riou, 2006, p 167): it is not an observable trait or a given, but an ideal or even assignment. The subject that is thus called into existence can accordingly be regarded as 'un homme rêvé', 'a dreamt up man', as international law scholar Mireille Delmas-Marty (1999, p 107) writes. This humanizing fiction is also presupposed by the idea of equality. As historian Sieb Stuurman (2017, p 1) writes in his study *The Invention of Humanity: Equality and Cultural Difference in World History*:

> Common humanity and equality are not primeval facts that patiently await discovery. [...] That all the people in the world constitute a single community is not an empirical fact. What it means is that people can be represented as members of the meta-community of an "imagined humanity" [...] Whenever a sufficient number of people embrace universal ideas of common humanity, the limits of the thinkable are extended and new course of action appear on the horizon.

Accordingly, the idea of equality is to be understood as 'culturally significant similarity', Stuurman (2017, p 7) argues. This implies that 'people *are* not equal, but [...] can be *represented* as equals' (Stuurman, 2017, p 7).

It is this vital idea of equality, this shared 'imagined humanity' that transcends the biological and the genetic, which is, according to human rights documents, compromised by eugenic practices, for example, through attempts at 'improving' the human species. Such practices presuppose the existence or possibility of superior and inferior types of people. Or as UNESCO's IBC (2015, sub 111) explains the meaning of equality for these kinds of situations:

> The goal of enhancing individuals and the human species by engineering the genes related to some characteristics and traits [...] impinges upon the principle of respect for human dignity in several ways. It weakens the idea that the differences among human beings, regardless of the measure of their endowment, are exactly what the

> recognition of their equality presupposes and therefore protects. It introduces the risk of new forms of discrimination and stigmatization for those who cannot afford such enhancement or simply do not want to resort to it.

The last sentence from this quotation touches upon the *second dimension of equality* in the context of human rights frameworks for biogenetic developments: equality as equal or equitable access to the fruits of biogenetic science. Article 12 of the UDHGHR offers a clear expression of this second dimension of equality:

> Benefits from advances in biology, genetics and medicine, concerning the human genome, shall be made available to all, with due regard for the dignity and human rights of each individual.

Similarly, Article 3 of the Oviedo Convention calls upon states to take measures to provide 'equitable access to health care of appropriate quality'. In these two provisions, the human right to science[3] and the human right to health[4] are brought together with the principles of equality and equity.

Clearly, equality as equitable access (equality's second dimension) is very much connected to equality as a protection against discriminatory and eugenic practices (equality's first dimension): without equal or equitable access to health, there is a grave risk that social-economic disparities are translated into health disparities, which is at odds with the idea that people are of equal worth. Moreover, unequal access to reprogenetic technologies can have eugenic effects. As US biolaw scholar Judith Daar (2017, p XIV) argues:

> Barriers to assisted reproductive technologies deprive the less wealthy, less white, less traditional, and less able-bodied of the tools necessary to achieve biological parenthood. This exercise of procreative deprivation is eugenic to the core.

Yet, there are several differences in focus between the two dimensions of equality as applied to the context of biogenetics. Whereas equality in its first dimension seems rooted in 'a relational theory of equality', focusing on equal intrinsic worth and bringing about certain social relationships among people, the second dimension seems connected to a more 'distributive theory of equality' (see Anderson 1999, p 313 for this distinction), aiming to bring about certain patterns of distribution and giving rise to individual rights to access the fruits of biogenetic technologies. As such, the first dimension of equality is more concerned with collective interests and values, and the second dimension more with individual ones.

Another difference concerns the relation between equality and genetics. Whereas equality as a protection against discriminatory and eugenic practices suggests that equality can best be realized through *ignoring* genetic differences between people ('regardless of genetic characteristics'), equality as equitable access to the benefits of biogenetic science suggests the opposite idea, namely that equality can best be realized through *using* genetic information and insights. Similarly, in his article 'Legal conceptions of equality in the genomic age', US legal scholar Mark Rothstein, distinguishes between the more traditional 'genome-blind conception of equality (Rothstein, 2007, p 456) and 'a new conception of equality based on recognizing human genomic variations and respecting individual differences' (Rothstein, 2007, p 462).

This brings me to a third and final important difference between the two dimensions of equality: their underlying view of the human. The first concept invokes a counterfactual, idealist and fictional account of the human, based on which 'society should be structured *as if* everyone is exactly the same in their biology' (Harden, 2021, p 251; emphasis added). The second approach implies a more genetic understanding of the human and of egalitarianism, based on which 'society should be structured to work to the advantage of people who were least advantaged in the genetic lottery' (Harden, 2021, p 251).

Both dimensions of equality are of great importance for human rights approaches to human genetics and can generally be regarded as mutually reinforcing. Still, when it comes to human germline gene editing, it is the first dimension of equality – equality as a protection against discriminatory and eugenic practices – that has, so far, been most prominent in intergovernmental human rights instruments dealing with heritable genome editing. More specifically, this first dimension can be regarded as one of the main concerns underlying the prohibition on genetic modification of offspring. As already discussed, the IBC (2015, sub 107) argues that the alternative to a ban 'would be to jeopardize the inherent and therefore equal dignity of all human beings and renew eugenics, disguised as the fulfilment of the wish for a better, *improved* life.' In a similar vein, the prohibition in Article 13 of the Oviedo Convention reflects concerns about eugenics, and the underlying fear 'of intentional modification of the human genome so as to produce individuals or entire groups endowed with particular characteristics and required qualities' (Explanatory Report to the Oviedo Convention, sub 89).

However, the tide now seems to be turning. In recent debates about the broader issue of human gene editing, much attention has been paid to questions related to equitable access to genetic modification technologies. Striking examples are the WHO report on genome editing (WHO, 2021), the Statement from the Organising Committee of the Third International Summit on Human Genome Editing (2023) and the comments that followed in the wake of these publications (de Vries and Baylis, 2021; Sample and

Devlin, 2023; Ledford, 2023). What is being suggested by all of these documents is that 'equity and access need to be at the forefront of innovation in human genome editing' (see title of de Vries and Baylis, 2021). However, the question is whether this is only the case for somatic gene editing – which is not banned – with the aim of making non-heritable genetic alterations to existing patients, or also for the much more controversial issue of heritable genome editing. According to a growing number of authors, the value of equal access should also take center stage in human rights approaches to heritable genome editing.

Human germline gene editing in the name of equality

Within the 'novel human rights-based approach' (Kleiderman, Ravitsky, and Knoppers, 2019) to heritable genome editing that is currently being proposed, equal access is one of the central values. This shift can already be recognized in the Nuffield Council's report. In this, the authors not only express concerns about genetic discrimination or the marginalization in society of those who have been born with certain genetic conditions (equality's first dimension), but also about equal access to heritable genome editing:

> Justice concerns are likely to arise where access to benefits is unequally distributed, where the potential benefits and risks are distributed differently (so that some enjoy a greater proportion of the benefit while others bear more of the risk) and where this distribution is linked to the distribution of other goods (so that the effect is to compound existing inequalities or entrench advantages). This would be the case if, for example, the service were only available at a cost that would be prohibitive to some who might wish to use it. (Nuffield Council, 2018, sub 3.66)

These risks would indeed loom large if heritable genome editing became available in fertility clinics worldwide, and thereby became integrated in the commercial offer of global reproductive markets. In all likelihood, reproductive use of gene editing will be a costly procedure, at least initially. Under such circumstances, and in the light of existing social-economic disparities, there is a serious danger that heritable genome editing will deepen existing social-economic inequalities in society, creating a division in society between what is perceived as genetic *haves* and *have-nots*.

For some authors, these concerns are all the more reason to be critical of proposals to eventually lift the ban on human germline modification (Andorno et al, 2020; Baylis, 2019, p 67). Others, however, insist that these concerns serve to underline the need for a carefully designed governance framework, 'a prudent path forward', in order to make 'responsible uses' of this controversial technology possible which are open to all, and not just an elite (Baltimore et al, 2015). In this vein, the Nuffield Council, as discussed, propagates a regulatory instead of a prohibitive approach to heritable genome editing. A proper governance framework would be able to make sure that this technology is applied in a way that is respectful of the welfare of the future person and does not produce, or exacerbate, social inequalities (Nuffield Council, 2018, sub 5.16). From this perspective, genetic modification of offspring would be compatible with the principle of equality, as long as it is appropriately governed and regulated.

Some authors go further by suggesting that, from a human rights perspective, we have a *moral obligation* to develop a translational pathway to germline gene editing in order to make this technology safe for clinical use, as long as equitable access can be guaranteed. As bioethics and law scholars Kleiderman, Ravitsky, and Knoppers (2019, p 512) write:

> Based on considerations of the right to health and the right to science, one could argue that we have a *moral imperative* to pursue basic research that may eventually lead to treatment using [heritable genome editing]. The goal would be to decrease human suffering and/or improve quality of life, as well as to reduce the burden placed on families and the cost to the healthcare system. Indeed, as has been argued in the literature, [heritable genome editing] will initially be a costly intervention, which may create barriers to access. It will thus be essential to ensure equitable access for those who would benefit from it most.

Within this approach, the right to health and the right to science are interpreted in such a way as to support the claim that, from the perspective of human rights, individuals have the right to access heritable genome editing in order to eradicate serious genetic diseases. Moreover, based on this reading of the right to health and the right to science, there would be a moral obligation to develop heritable genome editing technologies.

Finally, there are authors who argue that not only the development and availability but also the *use* of heritable genome editing is morally imperative in the name of equality (Gyngell, Bowman-Smart, and Savulescu, 2019; Buchanan et al, 2000; Giubilini and Minerva, 2019). They regard genetic interventions as a more effective way to 'create' or 'enhance equality' (Giubilini and Minerva, 2019) than social-political ones. Indeed, according to them, heritable genome editing should be used to correct the unfair outcomes of the genetic lottery. Genetic factors, they point out, greatly affect the opportunities one has in many segments of life, ranging from education and work to health and love life. Thus, instead of addressing the resulting inequalities on a social level, one could choose to target the genetic roots of these inequalities by using germline modification.

In this vein, Christopher Gyngell, Hilary Bowman-Smart, and Julian Savulescu propose to promote equality in education in a near future, not so much through special educational programmes (Harden, 2021), but by using heritable genome editing, targeting individuals with low polygenic scores for educational attainment, thereby 'evening out the genetic starting point'. Such a measure 'could prove the most effective way' to address educational inequities 'with the additional benefit of being passed to future generations' (Gyngell et al, 2019, p 521). They thus conclude:

> The human genome was created by a blind process of mutation and selection occurring over thousands of generations. This process had no foresight for the creatures it would produce. This has resulted in vast natural inequality. [...] We ought to use powerful technologies like [heritable gene editing] to correct these inequalities and promote human flourishing. Such actions are *moral imperatives*. (Gyngell, Bowman-Smart, and Savulescu, 2019, p 522; emphasis added)

In this line of thinking, equality is also understood as a distributive issue. Not only is equitable access to heritable gene editing being supported in various ways; but ultimately, in these scenarios, combinations of genes for certain talents and traits are to be redistributed.

Shortcomings of the equal access approach from a human rights perspective

Is the conclusion, therefore, that human rights advocates and egalitarians should endorse human germline gene editing? Should equality, interpreted as equal or equitable access, be decisive in this context? Some have invoked John Rawls' *Theory of Justice* and his ideal of distributive justice in this context (Fox, 2007; Buchanan et al, 2000). In his legal-philosophical classic, Rawls briefly ponders the question of genetics. If genetic traits ever became redistributable, Rawls (1973, p 108) stipulates that people in the so-called 'original position', behind 'the veil of ignorance', would opt for the best genes for their offspring:

> [I]t is [...] in the interest of each to have greater natural assets. This enables him to pursue a preferred plan of life. In the original position, then, the parties want to insure *for their descendants the best genetic endowment*. [...] Thus over time a society is to take steps at least to preserve the general level of natural abilities and to prevent the diffusion of serious defects. [...] We might conjecture that in the long run, if there is an upper bound on ability, we would eventually reach a society with *the greatest equal liberty* the members of which *enjoy the greatest equal talent*. (emphasis added)

More than half a century later, germline gene editing *is* making it possible to change the genetic endowment of descendants. According to the authors of the book *From Chance to Choice: Genetics and Justice*, approaches that are based on Rawlsian egalitarianism and distributive justice are perfectly compatible with heritable genome editing. They stress that the rise of genetic technologies, such as heritable genome editing, will bring along a 'colonization of the natural by the just' (Buchanan et al, 2000, pp 82–84). When our power to control the biogenetic aspects of life expands, 'the territory of the natural is annexed to the social realm, and the new-won territory is colonized by ideas of justice' (Buchanan et al, 2000, p 84). As part of this colonization process, heritable genome editing could be regarded as a technology which makes it possible to apply the ideal of distributive justice to 'natural assets', that is, genes.

Against this background, recent calls to allow or encourage heritable genome editing in the name of equality may appear, at first sight, as egalitarian in nature and as required in a just society. However, I argue that if these calls can be called egalitarian, it is a very particular type of egalitarianism that radically differs from, and even clashes with, traditional human rights conceptions of equality, on at least four levels.

First, even when human germline gene editing is only used for medical, non-enhancement purposes, this technology does not cure patients or treat diseases, but aims to create new, healthy lives. Therefore, strictly speaking, germline interventions do not serve therapeutic purposes (Andorno et al, 2020, p 352; MacKellar, 2021) and do not 'treat or save lives of people that would otherwise have a genetic disease' (Rulli, 2019, p 1072). In that sense, it is also misleading to speak in terms of a right to be cured through germline gene editing, as part of the right to equal access, healthcare, or science. Moreover, undergoing treatment through germline gene editing cannot be seen as a human right, because no legal subject exists at that moment in time. Accordingly, the conclusion seems justified that:

> The real value of [heritable genome editing] is not to the individual who is created, but to the parents who choose to create the child. It might help them have a genetically related child free of a genetic disease. It does not save lives or treat anyone; rather it helps existing adults achieve their reproductive preferences. (Rulli, 2019, p 1078)

If one still chooses to regard access to heritable genome editing as a matter of justice, then it is not so much 'justice to persons', but 'justice in the designed creation of persons', as the authors of *From Chance to Choice* admit (Buchanan et al, 2000, p 85). In that process, they argue that it will no longer be possible to make a clear distinction between the subjects and objects of distributive justice, that is, between persons and goods (Buchanan et al, 2000, p 86). In that case, they write:

> [I]ntervening in the natural lottery by genetic means will not be limited to distributing goods among persons whose identities are fixed prior to the act of distribution. Instead, the distribution of genes itself may in part determine the identities of the persons. (Buchanan et al, 2000, p 85)

The first dimension of equality, which stresses the equal intrinsic worth of individuals as persons, suggests that such a blurring of the boundary between persons and things through the possible genetic design of future people, even for supposedly egalitarian purposes, is highly problematic. The fact that one person genetically designs another person, 'may give rise to a novel, curiously asymmetrical type of relationship between persons' (Habermas, 2003, p 42), 'a social relationship in which the usual reciprocity between persons of equal birth is revoked' (Habermas, 2003, p 64). It is for the same reason that the Oviedo Convention's Explanatory Report (sub 89) describes 'the intentional modification of the human genome so as to produce individuals or entire groups endowed with particular characteristics and required qualities' as 'the ultimate fear'. New inequalities threaten to take shape as soon as a distinction can be made in society between the subjects and 'objects' of genetic design, between the designers and those who have been designed.

Second, the focus on equal access and equity distracts from the collective question as to whether heritable genome editing should be permitted in the first place, a matter which requires broad public and societal engagement. By placing equity and access at the forefront of the discussion, one has tacitly moved from asking *whether* to proceed and whether that would be in line with the common good, to asking *how* to proceed with heritable genome editing (Hurlbut, 2019) and how individuals can be secured access. Viewed solely through the lens of equality as equitable access, it is tempting to regard the desirability of heritable genome editing as a matter that can be resolved on an individual level, as long as equitable access has been properly addressed. In that process, one loses sight of the bigger picture, as well as the collective values and interests that are at stake with allowing or prohibiting genetic modification of offspring through heritable genome editing. As Canadian bioethicist Françoise Baylis argues:

> We need time – time to agree on what are our shared problems and how heritable genome editing might or might not help address them. Instead, too often our attention is focused on individual goals and objectives. We fiddle while Rome burns. Instead of taking time (and eventually action) to address our shared needs, we debate whether to support the reproductive desires of hypothetical prospective parents

> who want to use heritable genome editing to have children with specific traits and have the resources to satisfy this "want". (Baylis, 2019, p 219)

Third, it appears that transhumanists, as advocates of human enhancement, also feel quite at home within an equal access approach to heritable genome editing. Some of them advocate giving access to germline gene editing to everyone, thereby allowing 'the ungifted to approach the gifted' (Savulescu, 2009, p 530). Others propagate giving access only to 'a certain class of disadvantaged individuals' (Giubilini and Minerva, 2019, p 336) and denying access to the more privileged classes. What both groups have in common, though, is that they regard genetic modification as an effective equalizer in society. From their perspective, the principle of equality can also justify using heritable genome editing for enhancement purposes, such as creating more intelligence or more strength in an individual.

As such, proposals to 'enhance equality' (see title of Giubilini and Minerva, 2019) through germline interventions border on the eugenic, as a distinction is being made between the winners and the losers of the genetic lottery, between the gifted and the ungifted, and between classes of genetically advantaged and disadvantaged individuals. In this vein, Gyngell, Bowman-Smart, and Savulescu (2019, p 520) propose that in order to create equality in education through the use of heritable genome editing, one will first need:

> [T]o stratify the population into three board groups for educational attainment "high predisposition to high intelligence"; "medium predisposition to high intelligence" and "low predisposition to high intelligence". It will become theoretically possible to use [heritable genome editing] to shift individuals from the low or medium predisposition groups, into the high predisposition group.

This indicates that an equal access approach to heritable genome editing can lead to results that are fundamentally at odds with existing human rights approaches to genetics. The Oviedo Convention and the UDHGHR oppose both eugenics and human enhancement based on equality's first dimension. As the IBC (2015, sub 111) writes: 'the goal of enhancing individuals and the human species by engineering the genes related to some characteristics and traits […] impinges upon the principle of respect for human dignity.'

Of course, not all advocates of an equal access approach to heritable genome editing will favour using this technology for enhancement purposes. For example, some propose making heritable changes only for serious genetic diseases (Kleiderman, Ravitsky, and Knoppers, 2019; NASEM/Royal Society, 2020). This approach has been chosen in many legal systems for the regulation of preimplantation genetic testing, a technology which allows the genetic selection of embryos *in vitro*. However, in the case of heritable genome editing, the boundary between therapy and enhancement will be much harder to maintain.

Because this technology is not only about genetic selection but also genetic modification, it breaks with the principle of recombination of parental genes. As a result, many more options, each with their own side-effects, trade-offs, and dilemmas, will become available, thereby blurring and eroding the boundary between healing and enhancing (van Beers, 2020, p 23).

This substantially increases the risk that the equal access approach will also result in the use of heritable genome editing for enhancement purposes. The He Jiankui case offers a striking illustration. Although the Chinese biophysicist aimed to make the baby girls in question immune to HIV, one of the unintended consequences of his interventions may have been to improve the girls' cognitive functions. In other words, cognitive enhancement (Regalado, 2019). The likelihood of such 'incidental enhancements' (Juengst, 2018, p 353) exemplifies the difficulty of keeping enhancement at bay once heritable genome editing is allowed.

Fourth and finally, the equal access approach to heritable genome editing suggests that equality is primarily a distributive matter, or even that equality can be created by 'levelling the playing field' through germline interventions. From the perspective of what could be called 'genetic distributive justice', equality is implicitly understood as having the right to the same 'natural assets' or the same 'genes' as others. But this view of equality is fundamentally at odds with the counterfactual, non-empirical view of humanity and dignity on which the human rights view of equality is premised. The idea is that, as Stuurman (2017, p 7) writes, 'people *are* not equal, but [...] can be *represented* as equals', based on their shared dignity. In this vein, within the Oviedo Convention and the UDHGHR, human dignity is understood as a principle which protects against genetic determinism. According to Article 2 of the UDHGHR, 'everyone has a right to respect for their dignity and for their rights *regardless* of their genetic characteristics' (sub a). Furthermore, 'that dignity makes it imperative not to reduce individuals to their genetic characteristics and to respect their uniqueness and diversity' (sub b).

In the attempt to create equality on a genetic level, invoking a supposed right to access heritable genome editing, these vital thoughts have been lost from view. More specifically, the danger is that a focus on equal access to germline modification or the equitable redistribution of genes may create the impression that the primary response to inegalitarianism lies in fixing 'bad genes' and changing the persons suffering from them, rather than in fixing social oppression and changing social attitudes towards people with genetic diseases or disabilities (as indicated in the chapter by Sufian and Garland-Thomson). Indeed, if heritable gene editing is ever allowed, and if this technology truly becomes available to all, as supported by those in favour of the equal access approach, it will become even harder to resist the thought that the problem lies on an individual genetic level, instead of being a collective responsibility of society.

Conclusion: 'equal in dignity'

Within the alternative human rights approach to human germline gene editing that is currently being proposed, the principle of human dignity is no longer regarded as 'the essential value to be upheld' (Explanatory Report to the Oviedo Convention, sub 9). In their discussion of heritable genome editing, some commentators (Mills, 2017; Alta Charo, 2018) have even gone as far as suggesting that it is possible to have human rights without an underlying notion of the human or of human dignity. Instead, the priority now appears to be on the principle of equality.

In this chapter, I have discussed the two main shapes that equality takes on in the context of the human rights governance of biogenetic developments: (1) equality as protection against discriminatory and eugenic practices, and (2) equality as equitable access to the benefits of biogenetic science. Normally, these two interconnected dimensions of equality mutually reinforce each other. Moreover, both dimensions are highly important in the collective process of deciding if there is a future for heritable genome editing, and if there is, in giving shape to that future.

However, in the recent proposals for a new human rights framework, equality as equal access is interpreted in such a radicalized way, that the result can be regarded as being at odds with equality in its first dimension on several levels. This suggests that the balance has been lost.

In order to do justice to the first dimension of equality, and to find a better balance between (1) equality as equal access and (2) equality as protection against discriminatory and eugenic practices, I suggest bringing the principle of human dignity back into focus. Moreover, in discussing what equality actually means for human germline gene editing, it will be important to answer the question concerning which view of the human is presupposed by equality. Evidently, dignity and humanity are concepts open to multiple interpretations. Still, one thing is clear. As Article 1 of the Universal Declaration of Human Rights makes clear, the underlying philosophy of human rights is not that we are all born equal in genes, but that we are born 'equal in dignity'.

Notes

1 For further reflection on Article 1 Universal Declaration on the Human Genome and Human Rights, see Jacobs' chapter in this volume.

2 'An intervention seeking to modify the human genome may only be undertaken for preventive, diagnostic or therapeutic purposes and only if its aim is not to introduce any modification in the genome of any descendants.'

[3] For example, Article 15 (1) sub b International Covenant on Economic, Social and Cultural Rights; Article 27(1) Universal Declaration of Human Rights.

[4] For example, Article 12 International Covenant on Economic, Social and Cultural Rights; Article 25 Universal Declaration of Human Rights.

References

Alta Charo, R. (2018) 'Germline engineering and human rights', *AJIL Unbound*, 112: 344–349.

Anderson, E. (1999) 'What is the point of equality?', *Ethics*, 109: 287–337.

Andorno, R. (2009) 'Human dignity and human rights as a common ground for a global bioethics', *Journal of Medicine and Philosophy*, 34(3): 223–240.

Andorno, R. et al (2020) 'Geneva statement on heritable human genome editing: The need for course correction', *Trends in Biotechnology*, 38(4): 351–354.

Baltimore, D. et al (2015) 'A prudent path forward for genomic engineering and germline gene modification', *Science*, 348(6230): 36–38.

Baylis, F. (2019) *Altered Inheritance: CRISPR and the Ethics of Human Genome Editing*, Cambridge, MA: Harvard University Press.

Boggio, A., Romano, C.P.R., and Almqvist, J. (2020) 'The governance of human (germline) genome modification at the international and transnational level' in A. Boggio, C.P.R. Romano, and J. Almqvist (eds), *Human Germline Genome Modification and the Right to Science: A Comparative Study of National Laws and Policies*, Cambridge: Cambridge University Press.

Boggio, A. and Yotova, R. (2021) 'Gene editing of human embryos is not contrary to human rights law: A reply to Drabiak', *Bioethics*, 35(9): 956–963.

Buchanan, A. et al (2000) *From Chance to Choice. Genetics and Justice*, Cambridge: Cambridge University Press.

Daar, J. (2017) *The New Eugenics: Selective Breeding in an Age of Reproductive Technologies*, New Haven: Yale University Press.

Delmas-Marty, M. (1999) 'Certitude et incertitudes du droit', in H. Atlan et al (eds), *Le Clonage Humain*, Le Seuil, pp 67–115.

de Miguel Beriain, I. (2018) 'Human dignity and gene editing', *EMBO Reports*, e46789.

de Vries, J. and Baylis, F. (2021) 'Equity and access need to be at the forefront of innovation in human genome editing', *The Conversation*, 12 July, Available from: https://theconversation.com/equity-and-access-need-to-be-at-the-forefront-of-innovation-in-human-genome-editing-161794.

de Wert, G. et al (2018) 'Responsible innovation in human germline gene editing: Background document to the recommendations of ESHG and ESHRE', *European Journal of Human Genetics*, 26(4): 450–470.

Explanatory report to the Convention on Human Rights and Biomedicine, European Treaty Series (1997), no. 164, Available from: https://rm.coe.int/16800ccde5.

Fox, D. (2007) 'Luck, genes, and equality', *Journal of Law, Medicine and Ethics*, 35(4): 712–726.

Giubilini, A. and Minerva, F. (2019) 'Enhancing equality', *Journal of Medicine and Philosophy*, 44(3): 335–354.

Gyngell, C., Bowman-Smart, H., and Savulescu, J. (2019) 'Moral reasons to edit the human genome: Picking up from the Nuffield Report', *Journal of Medical Ethics*, 45(8): 514–523.

Habermas, J. (2003) *The Future of Human Nature*, Cambridge: Polity Press.

Harden, K.P. (2021) *The Genetic Lottery. Why DNA Matters for Social Equality*, Princeton/Oxford: Princeton University Press.

Hurlbut, J.B. (2019) 'Human genome editing: Ask whether, not how', *Nature*, 565: 135.

IBC (International Bioethics Committee) (2015) *Report of the IBC on Updating Its Reflection on the Human Genome and Human Rights*, SHS/YES/IBC-22/15/2 REV.2, Paris.

Juengst, E.T. et al (2018) 'Is enhancement the price of prevention in human gene editing?', *CRISPR Journal*, 1(6): 351–354.

Kleiderman, E., Ravitsky, V. and Knoppers, B. (2019) 'The "serious" factor in germline modification', *Journal of Medical Ethics*, 45(8): 508–513.

Labrusse-Riou, C. (2006) 'L'Humain en Droit: Réalité, Fiction, Utopie?', in C. Hervé and J.J. Rozenberg (eds), *Vers la fin de l'homme?*, Brussels: De Boeck, pp 157–176.

Ledford, H. (2023) 'Beyond CRISPR babies: How human genome editing is moving on after scandal', *Nature*, 2 March, https://doi.org/10.1038/d41586–023-00625-w.

MacKellar, C. (2021) 'Why human germline genome editing is incompatible with equality in an inclusive society', *The New Bioethics*, 27(1): 19–29.

Mills, P. (2017) 'Genome editing, human rights and the 'posthuman', 3 October, Available from: http://nuffieldbioethics.org/blog/genome-editing-human-rights-posthuman.

NASEM (2017) *Human Genome Editing: Science, Ethics, and Governance*, Washington, DC: The National Academies Press, https://doi.org/10.17226/24623.

NASEM/Royal Society (2020) *Heritable Human Genome Editing*, Washington, DC: The National Academies Press, https://doi.org/10.17226/25665.

Nuffield Council (2018) 'Genome editing and human reproduction: social and ethical issues', Available from: https://s47055.pcdn.co/wp-content/uploads/Genome-editing-and-human-reproduction-report.pdf.

Organising Committee of the Third International Summit on Human Genome Editing (2023) 'Statement from the Organising Committee of the Third International Summit on Human Genome Editing', 8 March, Available from: https://royalsociety.org/news/2023/03/statement-third-international-summit-human-genome-editing/.

Rawls, J. (1973) *A Theory of Justice*, Oxford: Oxford University Press.

Regalado, A. (2019) 'China's CRISPR twins might have had their brains inadvertently enhanced', *MIT Technology Review*, 21 February.

Rothstein, M.A. (2007) 'Legal conceptions of equality in the genomic age', *Minnesota Journal of Law & Inequality*, 25(2): 429–463.

Rulli, T. (2019) 'Reproductive CRISPR does not cure disease', *Bioethics*, 33: 1072–1082.

Sample, I. and Devlin, H. (2023) 'Forthcoming genetic therapies raise serious ethical questions, experts warn', *Guardian*, 6 March, Available from: www.theguardian.com/science/2023/mar/06/forthcoming-genetic-therapies-serious-ethical-questions-experts

Savulescu, J. (2009) 'Genetic interventions and the ethics of enhancement of human beings', in B. Steinbock (ed), *The Oxford Handbook of Bioethics*, Oxford: Oxford University Press, pp 516–535.

Stuurman, S. (2017) *The Invention of Humanity: Equality and Cultural Difference in World History*, Cambridge: Harvard University Press.

van Beers, B.C. (2020) 'Rewriting the human genome, rewriting human rights law? Human rights, human dignity, and human germline modification in the CRISPR era', *Journal of Law and the Biosciences*, 7(1): lsaa006.

van Est, R. et al (2016) 'Rules for the digital human park; Two paradigmatic cases of breeding and taming human beings: Human germline editing and persuasive technology', Background Paper for the 11th Global Summit of National Ethics/Bioethics Committees, Berlin, 16–18 March 2016.

WHO Expert Advisory Committee on Developing Global Standards for Governance and Oversight of Human Genome Editing (2021), 'Human genome editing: a framework for governance'.

5

The Human Genome as Heritage of Humanity: Why Equality Matters

Laura E. Jacobs

Introduction

The United Nations Educational, Scientific and Cultural Organization (UNESCO) 1999 document on the *Birth of the Universal Declaration on the Human Genome and Human Rights* indicated:

> What human intervention produces, forms and recreates, dictates the future of all mankind, for as Einstein said: we shall have the destiny we deserve. This should be the inspiration for the continuity of the work of UNESCO's International Bioethics Committee (IBC). (UNESCO, 1999, p 41)

With these promising words, the French politician Noëlle Lenoir[1] concluded one of the 1993 preparatory reports for the drafting of the 1997 Universal Declaration of the Human Genome and Human Rights ('the UDHGHR'). In Article 1 of this Declaration, the human genome is declared to be the *heritage of humanity*. 'Heritage of humanity' is a unique concept, as it does not often feature in regulations (Ossorio, 2007, p 425). Moreover, it is the first time that 'man as such [...has] been regarded as a common heritage[2] to be protected' (Byk, 1998, p 235). Those who drafted the UDHGHR emphasized that the proclamation of Article 1 is the cornerstone of the Declaration as all following provisions proceed from the innovative nature of this first article (UNESCO, 1997b, p 8; 1999, p 107).

Notwithstanding the obvious importance of the first article, it appears that diverging ideas exist about what is meant by this 'heritage status' of the human genome. Some scholars believe that it is, primarily, a utilitarian concept demanding that genetic materials – like all of the world's resources – should be utilized for the greater good and benefit of all of humanity (Underkuffler, 2007, p 398; Gomes de Andrade, 2010, p 441), while others regard the heritage status of the human genome as a rationale for protecting the human genome against improper manipulation (Andorno, 2013, p 33). Given the increasing technological possibilities aimed at editing the human genome, such ambiguities are problematic, because the 'status' of the human genome is often considered in, and of great importance for, discussions about genetic modifications, such as heritable genome editing (Greely, 2021, pp 209–211). As the biophysicist Mara Almeida and the bioethicist Robert Ranisch put it: 'the modification of the human genome raises complex questions about the characterization of the human species genome and if there should be limits on interfering with it' (Almeida and Ranisch, 2022, p 8).

Limits on interfering with the human genome are currently in place, as bans on heritable genome editing exist in most countries and legal regulations around the world (Baylis et al, 2020). These bans have, however, initiated considerable debate now that several scientists suggest making heritable genome editing permissible – instead of banning it altogether – provided that certain criteria are satisfied (Baylis, 2019, pp 133–135). Since this debate is likely to persist (Macintosh, 2022; Baylis, 2023), it is of great importance to re-examine one of the key provisions on the human genome. Article 1 of the UDHGHR is such a key provision, because it distinctly conceptualizes the human genome by declaring it to be the heritage of humanity. Moreover, it is the only universal piece of legislation that governs the human genome so explicitly. However, it remains unclear what this conceptualization encompasses and what its implications are.

The aim of this chapter is, therefore, to look more closely into the meaning of the heritage status of the human genome. This will be done by thoroughly examining its legislative history. Although it will be demonstrated that the concept of 'heritage of humanity' has evolved over time, it will be argued that one notable underlying element has always been part of it: *the notion of equality*. It will further be asserted that in the context of genetic heritage, this notion of equality encompasses three different understandings: *equal access*, *equality among generations*, and *equal dignity*. It will also become clear that, depending on the context in which the heritage concept is applied, different understandings of the notion of equality are, to a greater or lesser extent, important. Finally, this chapter will maintain that, by means of its egalitarian nature, the heritage status of the human genome dictates protection of the human genome. In this way, it will be argued that the concept of the human genome as heritage of humanity is compatible with the previously described bans on heritable genome editing.

In order to make these arguments, the study will first analyse what the legal roots of 'common heritage of mankind' are, and how this concept was understood when it made its first entrance in international law. Second, the drafting reports of the UDHGHR will be examined in order to find out how the heritage concept in the context of the human genome came into being during its drafting process. Third, it will be argued that the inherent egalitarian nature of the human genome supports a focus on the protection of the human genome, which sustains the predominantly prohibitive approach to heritable genome editing worldwide. In the conclusion the arguments will be summarized.

Legal roots of 'common heritage of humanity'

The legal roots of the concept 'common heritage of humanity' are to be found in the Roman legal notion of *res communis*: the principle of making things a matter of common resource instead of individual ownership (Byk, 1998, p 238). Although both concepts have similarities, the common heritage emerged over time as a fairer alternative to *res communis*. Accordingly, as will be discussed, these concepts need to be distinguished from each other. The first time the concept of the common heritage of humanity was mentioned can be traced back to international law documents relating to the law of the sea (UNESCO, 1999, p 54). In this section, the development of the heritage concept as it arose in the context of the deep seabed will therefore be analysed.

Heritage and equal access

The 17th century's Dutch jurist Hugo Grotius, who is regarded as one of the founders of international law, considered the sea to be *res communis* (Buttigieg, 2018, p 20). In modern usage, matters *res communis* share two characteristics: first, they may not be appropriated as such and second, they belong equally to all people, hence, the matter is open to use by all members of the community (Larschan and Brennan, 1982, p 305). This means that within the *res communis* doctrine, it is possible to become the owner of a fruit reaped from a *res communis*. The fisherman who fishes in the sea is, according to this doctrine, allowed to keep the fish. Hence, *res communis* grants free access to resources beyond national jurisdictions. However, it does not prevent the situation that only those states who have the means to mine the deep seabed, would do so. Thus, *res communis* grants *free* access, but not *equal* access (Larschan and Brennan, 1982, p 317).

Since 1873, private companies of the wealthier countries started to claim exclusive mining rights on the deep seabed in the Pacific (Bovenberg, 2006). This made Arvid Pardo, a Maltese diplomat and international law professor, anxious about the likely monopolization of the resources in the deep sea by the wealthier countries (Ida, 1998, p 59). In his role as Malta's permanent representative to the United Nations, he held a memorable speech in 1967 to the First Committee of the United Nations General Assembly. In this speech, Pardo spoke for over three hours about the 'truly inestimable value' of the deep sea and sketched the hazard of technologically advanced states rushing to colonize the seabed, resulting in these treasures being the sole benefit of a few nations (Ranganathan, 2021, p 237). In order to prevent this from happening, he proposed to declare the deep seabed as the common heritage of humankind; an alternative to *res communis*.

The alternative of the common heritage of humanity concept was for Pardo, 'first and foremost, [...] the right to equal share' (Baslar, 1997, p 32). No single definition of the common heritage of humanity concept exists (Arnold, 1975, p 153), but from Pardo's following quote several features become apparent:

> The objective of my government has been as follows: preservation of the international character of the seabed and ocean floor [...], not as a res omnium communis, usable for any convenient purpose and the resources of which are indiscriminately and competitively exploitable, but through the acceptance by the international community of the principle that these vast areas of our planet have special status as a common heritage of mankind and, as such, should be reserved exclusively for peaceful purposes and administered by an international agency in the name and for the benefit of all peoples and of present and future generations. (Pardo, 1969, p 214)

It follows that the common heritage concept includes the following five elements: (1) the common heritage is not subject to national appropriation (thus can be used but not owned); (2) usage of the common heritage requires a balanced system of management; (3) benefits resulting from exploitation of the resources of the common heritage must be actively shared; (4) the common heritage shall be reserved exclusively for peaceful purposes; and (5) common heritage promises keeping the deep seabed in trust for future generations (Pardo, 1983).

Upon closer examination of these five elements, it becomes clear that *res communis* is indeed a distinct legal concept from the common heritage of humanity concept. This is the case because only the first element is shared with the *res communis* doctrine, while the others are not. Furthermore, it appears that almost all these elements stem from Pardo's core aim as reflected in his memorable speech, namely, ensuring *equal access*. Especially the first, second, third, and fourth element aim at ensuring equal access for all countries to the fruits of the deep seabed (Wolfrum, 1983, p 319).[3] From these four elements, one can hence infer the first understanding of equality within the heritage concept: Pardo's heritage concept aims to realize equality in the *understanding of equal access.*

Heritage and equality among generations

Not all scholars who write about the common heritage concept agree that the fifth element of reservation for future generations, sometimes also referred to as the 'intergenerational' scope, is part of the heritage concept (Sturges, 1999). Perhaps this can be explained by the fact that Pardo's own vision on what the heritage status was supposed to encompass, was broader than the finally adopted text in the law of the sea convention (Kuppuswamy, 2009). Some scholars argue that by choosing the word 'humanity', 'mankind', or 'humankind' (instead of, for example, 'states'), the intergenerational scope is undoubtedly ingrained in the common heritage concept (Dupuy, 1983, p 348; Wolfrum, 1983, pp 318–319). Namely, the interests of 'all mankind' are separate and distinct from those of all states – the notion of 'all mankind' thus encompasses people who are not incorporated in political entities and includes, for example, people who still have to be born (Buttigieg, 2018, p 16). Some Ukrainian legal scholars speak in this respect of humanity as a 'trans-historical bearer of rights and responsibilities', meaning that each generation should use the common heritage of mankind with the responsibility to past and future generations (Radzivill et al, 2020, p 158). Others see the focus on future generations as a logical implication of the equal access understanding, because it will lead to the avoidance of a 'tragedy of the commons' scenario: the situation in which unrestricted access to resources will ultimately lead to over-exploitation, leaving future generations with nothing left to equally share (Shackelford, 2009, pp 117–118).

In other domains in which the heritage concept is used, a focus on future generations can also be found. Following legislation on cultural heritage, codified in the UNESCO *1972 Convention concerning the Protection of the World Cultural and Natural Heritage*, it appears that states have a duty, in Article 4, 'of ensuring the identification, protection, conservation, presentation and transmission to future generations of the cultural and natural heritage situated on its territory'. Furthermore, in the *Agreement Governing the Activities of States on the Moon and Other Celestial Bodies*, in which the Moon is declared to be common heritage of humankind in Article 11, it is stated that due regard shall be paid to interests of present and future generations (United Nations, 1979). And also UNESCO's *1997 Declaration on the Responsibilities of the Present Generations Towards Future Generations*[4] indicates in Article 8 that present generations may use the common heritage of humankind as defined in international law provided that this does not entail compromising this heritage irreversibly. A condition for using the common heritage at the present time thus includes the consideration of the interests of future generations who should also be able to make use of it.

Accordingly, the heritage concept directly relates to the idea that living people have obligations to future generations. Thus, the heritage concept seeks to protect and conserve the interest of these future generations. This *notion of equality among generations* (sometimes referred to as 'intergenerational equity' or 'intergenerational justice') is deeply rooted in international law, especially in the context of the environment. In this area of law, equality among generations means that 'all generations are *equal* in their normative position in relation to the natural system of which they are a part. There is no basis for favoring one generation over another' (Brown Weiss, 1990).

The main purpose of declaring the deep seabed the common heritage of mankind was clearly to ensure equal access to the fruits of the deep sea for both present *and* future generations. As Pardo was concerned with the interests of future generations, it should be concluded that *equality among generations* is a second understanding of equality that, next to the first understanding of *equal access,* can be recognized in Pardo's heritage concept. In the next section, an analysis will be undertaken to see whether these two understandings of equality also underlie the heritage concept in the context of the human genome.

Development of the heritage concept in the UDHGHR

The human genome is obviously a very different subject compared to the deep sea. As a result, for the UDHGHR, the characterizations of the heritage concept within international law had to be 'adapted to the specificities of the human genome' (UNESCO, 1995b, p 107). This was first of all accomplished by declaring the human genome the heritage of humanity *in a symbolic sense.* Such wording was chosen to dispel the concerns of Member States, specifically that (1) the common heritage notion would be interpreted as 'common property' and could encourage appropriation of genetic sequences by companies and (2) the original wording would pose a threat to individual rights as it could be interpreted as a right to enhancing the human gene pool in favour of eugenic policies (Andorno, 2013, p 87). The phrase *in a symbolic sense* was hence used to express that 'the human genome should not be considered as being capable of some collective form of appropriation' (UNESCO, 1997a, p 9).

In this section the drafting reports of the UDHGHR will be analysed to further understand how the adaptation came about. This will reveal that heritage in the genetic context is both similar to and different from heritage in the context of the deep seabed. *Similar* because within genetic heritage both understandings of the notion of equality, equal access, and equality among generations, can be recognized. *Different* because in the genetic context, a third understanding of equality becomes apparent: equal dignity.

The UDHGHR and equal access

The first understanding of equality as *equal access,* also played a role within the heritage concept in the genetic context. This can, for example, be found in a report of the Director-General of UNESCO:

> Using the notion of the common heritage of humanity, found in international law, also lays stress on the notion of the equitable sharing among countries of the advantages it is hoped that its exploitation may provide. This is the view that predominates in the case of the seabed or outer space. The notion of the common heritage of humanity thus makes it possible to take into account the objective of ensuring fair exploitation of the results of research on the human genome. The aim is to guarantee the right of access of all people and of all countries, especially the developing countries, to such results. (UNESCO, 1995a, p 15)

Equality in the understanding of equal access can be recognized in this quote because the drafters mention that results of research on the human genome should be accessible to everyone and explicitly relate this to Pardo's heritage of the seabed. Also in the response to the international consultation of one of the drafts of the UDHGHR, it is stated that the heritage concept should serve as a reminder that knowledge about the human genome can never be the privilege of an élite but must benefit humanity as a whole (UNESCO, 1995a, p 15). In other words, the benefits must be equally accessible. And finally, the idea of equal access also aligns with Article 12(a) of the UDHGHR which states that: 'Benefits from advances in biology, genetics and medicine, concerning the human genome, shall be made available to all [...]'. These three examples show that equality in the understanding of equal access was also underlying the heritage concept in the genetic context.

The UDHGHR and equality among generations

In the introduction to the UDHGHR the drafters wrote that:

> [T]he idea [to assert that, in a symbolic sense, the human genome is the heritage of humanity] is to underline the fact that research on the human genome and the applications that flow therefrom entail the responsibility of humanity as a whole, in the interest of present and future generations. (UNESCO, 1999, p 3)

In addition, in the fourth meeting of the Legal Commission of the IBC it was stated that the wording 'humanity', refers to a 'legally applicable concept, humanity being regarded as a subject of international law, with rights and responsibilities towards itself and towards future generations' (UNESCO, 1999, p 54). Moreover, in a report by the then UNESCO's Director-General, Dr. Federico Mayor Zaragoza from Spain, it is indicated that the aim of declaring the human genome the heritage of humanity was 'above all [...] fundamentally as a form of capital which humanity has a duty to safeguard for future generations' (UNESCO, 1995a, p 15). Accordingly, the drafters of the UDHGHR included a duty of present generations to future generations; the duty to safeguard the human genome for those future generations. Mr. Bedjaoui, member of the Legal Commission of the IBC, formulated this endeavour eloquently during a speech to the IBC: 'every individual, becoming automatically, at conception, the owner of this genetic heritage [...] [is] also under an obligation to protect it and safeguard it, in a word to assume responsibility for it, in view of our obligations towards the generations to come' (UNESCO, 1995b, p 107).

Arguably, the heritage status of the human genome is also aimed at effectuating equality in the understanding of *equality among generations*, as was the case with Pardo's heritage in the context of the deep seabed. *Equality among generations* is plausibly even more relevant in the genetic context than in the context of the deep seabed. As genes are passed down between generations, clearly, each one of us inherits them from his or her forebearers. The human genome is said to unite each person with both his or her ancestors and descendants, and thereby to effectuate a connection among all human beings as members of the human family of which past, present, and future generations are part (Ossorio, 2007, p 425). In the words of the Maltese legal scholar Emmanuel Agius: 'nothing is so clearly heritable as the genes' (Agius, 1998, p 75).

The UDHGHR and equal dignity

When analysing the legislative history of the UDHGHR, it appears that a third understanding of equality formed an important rationale for declaring the human genome the heritage of humanity: *equal dignity*. It follows both from the UDHGHR's Article 1 itself, as from the legislative history, that (human) dignity underlies the heritage concept in the context of the genome. In this section, it will therefore be argued that dignity can be interpreted as a third understanding of the notion of equality, as equality and dignity are very much intertwined.

Already from the wording of Article 1 of the UDHGHR it appears that dignity plays an important part. Indeed, prior to declaring the human genome the heritage of humanity, it is declared that 'the human genome underlies the fundamental unity of all members of the human family, as well as the recognition of their inherent dignity and diversity'. In the drafting reports this becomes further apparent. For example, in the fifth meeting of the Legal Commission of the IBC, it was argued that the term "common heritage of humanity', which was attributed to the human genome, should [...] be maintained in the [...] view of the need to guarantee respect for human dignity and human rights [...]'. Furthermore, the drafters asserted that 'the protection of the human genome is to safeguard [...] the dignity of each of the members of the human family, in other words, each individual' (UNESCO, 1999, p 47). This latter quote strengthens the view that the concepts of 'equality' and '(human) dignity' are very much intertwined,[5] because the UDHGHR demands recognition of the inherent dignity of *each of the members of the human family.* It thus confers upon each human being an equal status. Human dignity is, accordingly, not graduated by features or characteristics that human beings have (Düwell, 2011, p 218).

This strong connection between the two concepts of equality and dignity is also evident in Article 1 of the Universal Declaration on Human Rights, which asserts that 'all human beings are born free and equal in dignity and rights'. In the UDHR, 'dignity' was among the drafters understood as a reference to the 'distinctive value or worth of, and the respect to, every human being' (Hughes, 2011, p 5). Human dignity demands equal treatment of every individual, because every individual has inherent worth and is intrinsically equal to all others. As a consequence, the notion of equality upholds human dignity. By using the wording 'equal dignity' the interwovenness of these concepts is thus expressed.[6]

In conclusion, the argument made in this section is that three understandings of equality are underlying the heritage concept in the context of the human genome. These are not only the two understandings of equality as recognized in Pardo's heritage concept, namely *equal access* and *equality among generations*, but also a third understanding, namely *equal dignity.*

The human genome as heritage of humanity and equality

Notwithstanding the fact that the three above-mentioned understandings of equality are all underlying the heritage concept in the context of the human genome, they are not given the same weight in the legislative history of the UDHGHR. In the following subsection it will be argued that equality in the understandings of *equality among generations* and *equal dignity* appear to be more prominent in the legislative reports of the UDHGHR than equality in the understanding of *equal access.*

Equality among generations and equal dignity are predominant in the UDHGHR

One of the drafting reports of the UDHGHR includes the following statement:

> The use of [the wording 'heritage of humanity'] is aimed *above all* at describing the human genome, the biological foundation of the identity of the human species, fundamentally as a form of capital which humanity has a duty to safeguard for future generations. *Above all*, it is the principle of human dignity that is the basis of the legitimacy of this requirement to safeguard the genome: it must be protected against manipulations which may violate this dignity […]. Using the notion of the common heritage of humanity, found in international law, *also* lays stress on the notion of the equitable sharing among countries of the advantages it is hoped that its exploitation may provide. This is the view that predominates in the case of the seabed or outer space. (UNESCO, 1995a, p 15) [italics added for emphasis]

From this quote it appears from the words 'above all' that safeguarding the human genome for future generations (*equality among generations*) and human dignity (*equal dignity*) are considered to be the most important reasons for declaring the human genome the heritage of humanity. Equitable sharing (*equal access*), as predominant in Pardo's heritage, appears – from the use of the word 'also' – to be of secondary importance. Moreover, the fact that the drafters explicitly mention that equal access predominated in Pardo's heritage, seems to suggest that the same does not apply for the UDHGHR. This shift of emphasis on different understandings of the notion of equality is also confirmed in other passages of the UDHGHR and its legislative history. First, Article 10 of the Declaration asserts that 'no research applications concerning the human genome should prevail over respect for the human rights, fundamental freedoms and human dignity'. As the equal access understanding of equality in the genetic context implies that research applications of the human genome should be equally shared, this quote can be interpreted as prioritizing equality in the understanding of equal dignity over equality in the understanding of equal access. Second, in the fourth meeting of the Legal Commission of the IBC, it was considered that the heritage status is evolving from a concept aimed at preventing excesses of sovereignty over spaces of communal interests to 'an element in the protection of the collective interests of humankind against possible infringements of human dignity affecting its common future' (UNESCO, 1999, p 55). Apparently, the drafters were aware that they were adding the aim of equal dignity to Pardo's original heritage concept.

When analysing the development of the meaning of the heritage of humanity concept, the following thus stands out: (1) the focus on future generations (*equality among generations*) has always been part of the heritage concept; (2) whereas equality in the understanding of *equal access* was predominant in Pardo's heritage, this understanding has somewhat receded into the background within the heritage concept in the genetic context; (3) whereas equality in the understanding *of equal dignity* has not become explicitly visible within Pardo's heritage,[7] this understanding of equality naturally becomes an important understanding of equality for the heritage concept in the context of the human genome. Hence, within the development of the heritage concept from the deep seabed to the human genome, *the emphasis of equality has shifted from equality in the understanding of equal access to equality in the understanding of equal dignity*.

Particularly complex type of heritage

What presumably explains this change in emphasis is that the human genome is an utterly different type of heritage than, for example, the deep seabed or the moon. This is because the human genome is both the individual heritage of each individual and the shared heritage of the human species (Sass, 1998, p 228). Human genomes reside in every individual and are unique to each individual as they contain important information about that individual, though altogether the human genome also embodies all of the people of the human species who exist and who will exist (Ossorio, 2007, pp 431–432). Hence, the fact remains that 'the individual's "humanity" is shared with all other individuals' (UNESCO, 1995b, p 107). As the Uruguayan diplomat Héctor Gross Espiel wrote in the Introduction to the UDHGHR: 'the concept of the human genome refers to all the genes of each individual [...] as well as to all the genes that constitute the human species' (UNESCO, 1999, p 2).

The circumstance that the human genome unequivocally exists in each and every one of us sets it apart from any other form of heritage. Each human being can be considered as the 'depository' both of his or her genome as well as the genome of the entire species (UNESCO, 1999, p 47). The special nature of the genome, the fact that it is not some externally accessible good, but that man *as such* – both collectively as a species and individually as every one of us on its own – has been declared the heritage of humanity, therefore explains that the emphasis on equal access has shifted to an emphasis on equal dignity. The heritage status of the human genome represents, as eloquently articulated by the Maltese legal scholar Jean Buttigieg, 'a reality that transcends its immediate existence as a collection of human genes', because 'it embodies our very humanness that is a safeguard of both our individual and species-identity' (Buttigieg, 2018, p 163). Arguably, the phrase *in a symbolic sense* of Article 1 of the UDHGHR can be interpreted as affirming this transcendence of immediate existence as a collection of human genes. The human genome is considered to be the heritage of humanity because its protection aims for something that goes beyond protecting that sole collection of human genes; it aims to safeguard equal dignity for all present and future human beings. The UDHGHR accordingly tries to protect the human genome out of fear that technological developments will violate the equal dignity of all human beings. According to Article 24 of the UDHGHR, heritable genome editing is one such technique that could violate the dignity of all human beings. Choosing to modify the genes of a future person in order to make that person a 'better' or 'healthier' person could increase discrimination against those considered genetically disadvantaged (Article 24 of the UDHGHR; Andorno et al, 2020). It could thus undermine the intrinsic worth of individuals who do not conform to a perceived standard. The IBC confirmed this in 2015, when it published the 'Report of the IBC on Updating Its Reflection on the Human Genome and Human Rights' (UNESCO, 2015). Within this report, the following is written about the heritage status of the human genome:

> According to UNESCO's definition of the world heritage, generally applied both to monuments and natural sites, this is to underline the outstanding value of *what should be protected and transmitted to future generations*. Nature is often understood as a limit to human freedom. At least in this case, building exactly on Article 1 of the UDHGHR, the argument is made that it should be rather considered as its premise, so that interventions on the human genome should be admitted only for preventive, diagnostic or therapeutic reasons and without enacting modifications for descendants, as affirmed in Article 13 of the Oviedo Convention. The alternative would be *to jeopardize the inherent and therefore equal dignity* of all human beings and renew eugenics, disguised as the fulfilment of the wish for a better, improved life. (UNESCO, 2015, p 26) [italics added for emphasis]

When using heritable genome editing to edit out certain conditions or characteristics, it could be argued that one will implicitly assign greater worth to a life without that condition or characteristic. By attributing desirability to only certain kinds of human existence, a judgement is conveyed regarding which lives are deemed more or less valuable.[8] Furthermore, equal dignity can be undermined when future persons are shaped according to certain standards or ideals of the current generation. This is not to say, however, as exemplified in both Article 3 of the UDHGHR and in its legislative documents, that this protection aims at ensuring the intangibility of the human genome, as the human genome 'by its nature evolves', and is 'subject to mutations' (UNESCO, 1995a, p 15, Article 3 of the UDHGHR). It is to say that the heritage status of the human genome reflects a symbolic concept that extends beyond simply being a collection of human genes, in order to protect the equal dignity of all present and future human beings.

Protection of the human genome

Consequently, because the heritage concept in the genetic context confirms the notion of equality predominantly in the understanding of equal dignity of all present and future human beings, the aim of UDHGHR is thus to *protect* the human genome. As heritable genome editing is likely to violate the equal dignity of present and future human beings, the heritage concept of the UDHGHR aligns with the prohibitive approach of heritable genome editing as recognized in international law and in most countries in the world (Article 13 of the Oviedo Convention; Baylis et al, 2020). The IBC explicitly confirmed this stance in its 2015 Report when it asserted that interventions on the human genome are not admitted when modifications for descendants would be enacted, as affirmed in Article 13 of the Oviedo Convention (UNESCO, 2015, p 26).

Some scholars have described the heritage concept as a utilitarian concept, interpreting it as a principle of the greatest good for the greatest number and believing that common heritage arguments are a manner of falsely assuming the superiority of numbers. For example, the Portuguese legal scholar Norberto Nuno Gomes de Andrade does so, by translating the heritage status of the genome to 'the idea that genetic materials – like all of the world's resources – should be utilized to provide the greater good and benefit to all of humanity, as to do less would mean failing to recognize the common interests and needs of all humanity' (Gomes de Andrade, 2010, p 441). However, this interpretation of the heritage concept does not align with the *egalitarian nature* of the heritage concept. The heritage concept is not about needing to *utilize* genetic materials to provide the greater good but rather the opposite; to *protect* genetic materials for present and future generations. If the heritage concept of the human genome is inextricably intertwined with the notion of equality and predominantly in the understanding of equal dignity, this implies that genetic materials should not be used to provide the greater good. To do so would inevitably violate the notion of equality as it would lead to diminishing every individual's equal worth. The fact that, as discussed in the section 'Legal roots of 'common heritage of humanity'', the common heritage doctrine differs from the common resource doctrine, further affirms the idea that the heritage concept cannot be understood as a utilitarian concept. The Canadian bioethicist and physician Michelle Bayefsky explains this coherently by indicating the following:

> Both frameworks [of common heritage and common resource] establish a public interest in the human genome, but the common heritage framework can be used to justify policies aimed at preserving and protecting the genome, while the common resource framework can be employed to justify policies for utilizing the genome for the public benefit. (Bayefsky, 2017, p 210)

In summary, this section argued that, whereas *equality among generations* has always been part of the heritage concept, in the genetic context the emphasis of equality has shifted from equality in the understanding of *equal access* to equality in the understanding of *equal dignity*. It has been argued that this change of emphasis can be explained by the fact that heritable genome editing could endanger the equal dignity of present and future human beings and that this requires protection of the human genome. Accordingly, the heritage status of the human genome sustains the legal prohibitions on heritable genome editing.

Conclusion

This chapter sought to more closely examine the meaning of the heritage status of the human genome. Indeed, as the status of the human genome is of great importance for the ongoing discussions on heritable genome editing, it is imperative to better fathom one of the key provisions on the human genome as defined in Article 1 of the UDHGHR. It has been argued in this chapter that the concept of heritage of humanity is inextricably intertwined with the notion of equality. By analysing the legislative history of the heritage concept (from Arvid Pardo's preliminary ideas to its application to the human genome in the UDHGHR), three understandings of equality can be recognized in the heritage concept as applied to the human genome. These three understandings are: equal access, equality among generations, and equal dignity. The context in which the heritage concept is applied, determines the relative importance of the different understandings of equality. In the genetic context, it has become apparent that the emphasis of equality is on the understandings of equality among generations and equal dignity. It is therefore argued that the core of the UDHGHR's heritage concept lies in equality in the understanding of equal dignity of present *and* future human beings. It is for this reason that the heritage concept demands protection of the human genome. The human genome as heritage of humanity, therefore, endorses the prohibitive approach on heritable genome editing as recognizable in international and national regulations, because heritable genome editing could violate the equal dignity of present and future generations. Despite the considerable time that has elapsed since the inception of the UDHGHR drafting process and the emergence of substantial advancements in biotechnologies, it may be argued that the UDHGHR is still of great significance for discussions on biotechnologies such as heritable genome editing. Albert Einstein's words served indeed as a great inspiration; protecting our shared genetic heritage will hopefully contribute to a dignified destiny for humankind, a destiny that all of us deserve.

Notes

1 Ms. Lenoir was the President of UNESCO's International Bioethics Committee (IBC), this Committee was entrusted with drafting the Universal Declaration on the Human Genome and Human Rights.

2 Throughout the literature the concepts 'common heritage of humanity', 'heritage of humanity', 'common heritage of mankind', and 'common heritage of humankind' are somewhat interchangeably used. Whereas the drafting reports of the UDHGHR mostly refer to 'common heritage of humanity', the final version of the UDHGHR speaks of 'heritage of humanity, in a symbolic sense', see subsection 'Development of the heritage concept in the UDHGHR' for more on this. Although in this chapter it will be mainly referred to 'heritage of humanity' due to the fact that this is the final wording used in the UDHGHR, references to those equivalents will also occur. This is because the concept originates from the Law of the Sea Convention in which those equivalents are used.

3 As for the fourth element this may not seem so at first glance, but legal scholar Wolfrum explained clearly that the seabed should be used for peaceful purposes only because 'it was argued that as the utilization of the sea-bed should be carried out for the benefit of all mankind and that since any military use, from its very nature, only served national interests, all military activities should be prohibited'. It is for that reason that also the fourth element aims at providing equal access.

4 For further reflection on the Declaration on the Responsibilities of the Present Generations Towards Future Generations, see Bellver's chapter in this volume.

5 For further reflection on the connection on the notions of equality and dignity, see also van Beers' chapter in this volume.

6 This wording is also used in *Report of the IBC on Updating Its Reflection on the Human Genome and Human Rights* (2015). This will be further elaborated in section 'Equality among generations and equal dignity are predominant in the UDHGHR' of this chapter.

7 Although Pardo's heritage concept was unmistakably aimed at effectuating equal access of the fruits of the sea by all people in the present and in the future, it could be the case that equality in the sense of equal dignity was underlying his thought. Because all people in the present and future world are inherently equal, they have an equal right to access these fruits. Allegedly, equality in the understanding of equal access flows from equality in the understanding of equal dignity. For further reflection on the connection between these two understandings of equality, see also van Beers' chapter in this volume.

8 For further reflection on how heritable genome editing could deny the equal moral worth of all human beings, see also Matthews' chapter in this volume.

References

Agius, E. (1998) 'Patenting life: our responsibilities to present and future generations', in E. Agius et al (eds), *Germ-Line Intervention and Our Responsibilities to Future Generations*, Springer, pp 67–84.

Almeida, M. and Ranisch, R. (2022) 'Beyond safety: Mapping the ethical debate on heritable genome editing interventions', *Humanities and Social Sciences Communications*, 9(1): 1–14.

Andorno, R. (2013) *Principles of International Biolaw: Seeking Common Ground at the Intersection of Bioethics and Human Rights*, Bruxelles: Bruylant.

Andorno, R. et al (2020) 'Geneva statement on heritable human genome editing: The need for course correction', *Trends in Biotechnology*, 38(4): 351–354.

Arnold, R.P. (1975) 'The common heritage of mankind as a legal concept', *The International Lawyer*, 9(1): 153–158.

Baslar, K. (1997) *The Concept of the Common Heritage of Mankind in International Law*, Leiden: Brill Nijhoff.

Bayefsky, M.J. (2017) 'The human genome as public: Justifications and implications', *Bioethics*, 31(3): 209–219.

Baylis, F. (2019) *Altered Inheritance: CRISPR and the Ethics of Human Genome Editing*, Cambridge, MA: Harvard University Press.

Baylis, F. et al (2020) 'Human germ line and heritable genome editing: The global policy landscape', *The CRISPR Journal*, 3(5): 365–377.

Baylis, F. (2023) 'Heritable human genome editing is "currently not permitted", but it is no longer "prohibited": So says the ISSCR', *Journal of Medical Ethics*, 49(5): 319–321.

Bovenberg, J. (2006) 'Mining the common heritage of our DNA: Lessons learned from Grotius and Pardo', *Duke Law & Technology Review*, 5(1): 1–20.

Brown Weiss, E. (1990) 'Intergenerational equity and rights of future generations', Available from: https://archivos.juridicas.unam.mx/www/bjv/libros/4/1985/11.pdf.

Buttigieg, J. (2018) *The Human Genome as Common Heritage of Mankind*, Stuttgart: ibidem-Verlag.

Byk, C. (1998) 'A map to a new treasure island: The human genome and the concept of common heritage', *Journal of Medicine and Philosophy*, 23(3): 234–246.

Dupuy, R.-J. (1983) 'The notion of the common heritage of mankind applied to the seabed space law', *Annals of Air and Space Law*, 8: 347–356.

Düwell, M. (2011) 'Human dignity and human rights', in P. Kaufmann et al (eds) *Humiliation, Degradation, Dehumanization*, Dordrecht: Springer Netherlands, pp 215–230.

Gomes de Andrade, N.N. (2010) 'Human genetic manipulation and the right to identity: The contradictions of human rights law in regulating the human genome', *SCRIPT-ed*, 7(3): 429–452.

Greely, H.T. (2021) *CRISPR People: The Science and Ethics of Editing Human*, Cambridge, MA: The MIT Press.

Hughes, G. (2011) 'The concept of dignity in the universal declaration of human rights', *The Journal of Religious Ethics*, 39(1): 1–24.

Ida, R. (1998) 'Human genome as common heritage of humankind', in N. Fujiki and D. Macer (eds), *Bioethics in Asia*, Eubios Ethics Institute.

Kuppuswamy, C. (2009) *The International Legal Governance of the Human Genome*, London; New York, NY: Routledge.

Larschan, B. and Brennan, B.C. (1982) 'Common heritage of mankind principle in international law', *Columbia Journal of Transnational Law*, 21(2): 305–338.

Macintosh, K.L. (2022) 'Heritable genome editing and cognitive biases: why broad societal consensus is the wrong standard for moving forward', *Journal of Law and the Biosciences*, 9(1): lsac002.

Ossorio, P.N. (2007) 'The human genome as common heritage: common sense or legal nonsense?', *The Journal of Law, Medicine & Ethics: A Journal of the American Society of Law, Medicine & Ethics*, 35(3): 425–439.

Oviedo Convention (1997) *The Convention for the Protection of Human Rights and Dignity of the Human Being with regard to the Application of Biology and Medicine.*

Pardo, A. (1969) 'An international regime for the deep seabed: Developing law or developing anarchy', *Texas International Law Forum*, 5(2): 204–217.

Pardo, A. (1983) 'Before and after the law of the sea – where now', *Law and Contemporary Problems*, 46(2): 95–106.

Radzivill, O. et al (2020) 'International legal and philosophical aspects of the new protection concept of the common heritage of mankind: Philosophical and legal worldview', *WISDOM*, 15(2):153–174.

Ranganathan, S. (2021) 'What if Arvid Pardo had not made his famous speech? (False) contingency in the making of the law of the sea', in I. Venzke and K.J. Heller (eds), *Contingency in International Law: On the Possibility of Different Legal Histories*, Oxford: Oxford University Press, pp 231–245.

Sass, H.-M. (1998) 'Introduction: Why protect the human genome?', *The Journal of Medicine and Philosophy: A Forum for Bioethics and Philosophy of Medicine*, 23(3): 227–233.

Shackelford, S. (2009) 'The tragedy of the common heritage of mankind', *Stanford Environmental Law Journal*, 27: 101–157.

Sturges, M. (1999) 'Who should hold property rights to the human genome? An application of the common heritage of humankind', *American University International Law Review*, 13(1): 219–261.

Underkuffler, L.S. (2007) 'Human genetics studies: The case for group rights', *The Journal of Law, Medicine & Ethics*, 35(3): 383–395.

UNESCO (1995a) *General Conference Twenty-eight Session, Report by the Director-General on the Possibility of Drawing up an International Instrument on the Protection of the Human Genome [28 C/38]*, Available from: https://unesdoc.unesco.org/ark:/48223/pf0000101328.

UNESCO (1995b) *Proceedings 1995 of International Bioethics Committee of UNESCO (IBC) Volume 1*, Available from: https://wayback.archive-it.org/10611/20170524061813/http://www.unesco.org/new/fileadmin/MULTIMEDIA/HQ/SHS/pdf/2VolumeI_en.pdf.

UNESCO (1997a) *UNESCO Final Report of Committee of Governmental Experts For the Finalization of a Declaration on the Human Genome [BIO-97/CONF.201/9]*, Available from: https://unesdoc.unesco.org/ark:/48223/pf0000109141.

UNESCO (1997b) *UNESCO Presentation of the 'Revised Preliminary Draft of a Universal Declaration on the Human Genome and Human Rights' [BIO-97/CONF.201/4]*, Available from: https://unesdoc.unesco.org/ark:/48223/pf0000111620.

UNESCO (1999) *Birth of the Universal Declaration on the Human Genome and Human Rights, UNESCO*, Available from: https://unesdoc.unesco.org/ark:/48223/pf0000119390.

UNESCO (2015) *Report of the IBC on Updating Its Reflection on the Human Genome and Human Rights*. SHS/YES/IBC-22/15/2 REV.2, Available from: https://unesdoc.unesco.org/ark:/48223/pf0000233258.

United Nations (1979) *Agreement Governing the Activities of States on the Moon and Other Celestial Bodies*.

Wolfrum, R. (1983) 'The principle of the common heritage of mankind', *Max-Planck-Institut für ausländisches öffentliches Recht und Völkerrech*: 312–337.

PART II

Would Heritable Genome Editing Undermine Equality in Society?

6

Does a 'Life Unworthy of Life' Actually Exist in the Case of Very Severe Genetic Disability?

Pia Matthews

Introduction

Heritable genome editing offers the possibility of changing the DNA of an individual and, in effect, reducing the number of babies born with a disability. Rather than regarding genetic conditions as part of human diversity, genome editing seeks to eliminate certain conditions on the basis that disability in itself is a harm. Indeed, heritable genome editing may 'propagate the view that a disabled person's life is not worth living' (Nuffield Council on Bioethics, 2018, p 9). Alongside heritable genome editing, the Nuffield Council lists preimplantation and prenatal testing followed by non-transference of an affected embryo or termination of the pregnancy, as 'current reproductive options' for dealing with undesirable genetic variants (p 5). This is one reason why it is important to ask whether a 'life unworthy of life' can actually exist in the case of very severe genetic disability.

'Staying alive is a lot of work for a disabled person in an ableist society' says Alice Wong in the introduction to her edited collection of accounts written by people with disabilities (2020, p xv). Wong's aim is not merely to keep on saying 'Hey, pay attention to us. We're here! We exist! We're just as human as you!' (p xxii). Rather, her collected stories contribute to the continuing challenge to ableism, which is discrimination in favour of non-disabled people. Ableism leads to the assumption that a life of disability in itself is a life unworthy of life (UN, 2021); disablism is discrimination against disabled people. A corollary to ableism is the view that disability in itself is always a harm or negative value often to the extent that non-existence is preferable to a life of disability. I write this chapter from the perspective of someone whose daughter has profound multiple and complex disabilities associated with Rett syndrome. The title of this chapter could invite a personal reflection on our lived experience. But a question framed in terms of whether a life unworthy of life can actually exist in the case of very severe genetic disability does not look towards good stories, even if some people with severe disabilities could articulate that experience. Rather, the question goes back to Wong's statement about existence: 'we exist! We're just as human as you!' Life unworthy of life questions moral status and the very existence of people with severe genetic disability. If life is unworthy of life, then non-existence seems to be preferable. This same argument can be used for selecting against a possible future life, that is ending or bringing about the non-existence of a life which already exists.

A life unworthy of life expressed as preference for non-existence is often related to the best interests of a disabled individual or the avoidance of the 'harm' of disability. However, the worthwhileness of life is not a medical decision (Singer and Kuhse, 1985; Anscombe, 2005; Perry, 2008; Wilkinson and Savulescu, 2018). Furthermore, some argue that the 'best interests' principle does not apply since the profoundly disabled have no interests; nor is there harm to be avoided since there is no person to be harmed in profound disability (McMahan, 2002). Thus, an evaluation of a life unworthy of life comprises many factors, of which the most significant are notions of equality, discrimination, perceptions of what constitutes severe disability, attitudes to disability, ableism, the perception that death is preferable to disability, and, not least, anthropological assumptions about persons. None of these factors have simple or straightforward application and all are the subject of contention. This chapter begins by outlining the assumptions that underpin some interpretations of so-called unworthy lives and of very severe genetic disability before detailing the shift towards associating worthwhileness of life with the assumed harm of disability that makes death preferable. The chapter then interrogates ableist discriminatory attitudes towards disability that not only deny the equal moral worth of all human beings but also challenge what it means to be a human person. If, as the evidence shows, disabled people 'are still being treated as second-class citizens' in all major areas of life including life-sustaining treatment decisions (Equality and Human Rights Commission, 2017, Foreword), then designating a life as unworthy of life in the case of very severe genetic disability is a matter of grave injustice.

Less worthy lives to unworthy lives

Debates on disability are often framed around the notion that some lives are of less worth than others. This notion is regarded as a matter of fact by some, a matter of discrimination by others. In *Unspeakable Conversations* (2003) the disabled disability rights lawyer, Harriet McBryde Johnson recalled one such debate she had with the philosopher Peter Singer. Johnson took issue with Singer's matter of fact view that since disability makes an individual worse off, it is better for infants, parents, and society if infants with severe disabilities are killed. For Johnson this is discrimination, moreover, discrimination under the guise of beneficence: the infant is better off dead, and everyone else is also better off as a result. In their book *Should the Baby Live?* Singer and Helga Kuhse claim that the notion of equality of human lives is rooted in what they regard as the discredited traditional sanctity of life doctrine which holds that life is valuable in itself, and not because of the qualities it may possess. Pointing to cases where clinicians could have prolonged the lives of certain disabled infants but decided not to, they then claim that, in practice, this rhetoric of equality is not taken seriously. The authors observe that these decisions were not medical decisions but rather decisions based on the judgement that certain disabilities make the individual's life not worth living. Using utilitarian logic, they argue that having a child with severe disabilities is likely to prevent the parents from going on to have a 'normal' child. Therefore, it would be better for the infant, parents, and society if the child does not continue to live either by withdrawal of treatment or active euthanasia (1985). Notably, the extended title of their book is *The Problem of Handicapped Infants*: disabled infants are characterized as 'problems'. The Singer–Johnson debate and the Singer and Kuhse's shift from the concept of a less worthwhile life to a life not worth living indicates that the question 'does a "life unworthy of life" actually exist in the case of very severe genetic disability' is not a question about whether some lives are less worth living. Rather, the question is asking whether some lives are not worth living *at all*. Despite Singer's protestation that Johnson should not feel threatened since, she says, 'he insists he doesn't want to kill me', Johnson is not reassured by attitudes that hold it is better if certain people did not exist and that killing them is a viable option (2003).

Very severe genetic disability

The qualifier in the question that names severe genetic disability indicates that the disability itself makes life unworthy, not the suffering (or lack of suffering) or actual condition of the individual. Notably, there is no clear definition of what constitutes a very severe genetic disability. Some genetic disorders such as Tay-Sachs, Batten disease, infantile onset encephalomyopathic mitochondrial DNA depletion syndrome (MDDS), and cystic fibrosis may be fatal. According to Dan Brock, Tay-Sachs is 'so devastating as to make a life not worth living;' failure to prevent the conception or birth of an individual with Tay-Sachs involves 'potential wrongful life or wrongful birth', but since the condition is 'rarely compatible with life' this does not become an issue for practice or public policy (2001, p 1).

Commenting on the legal case of the English child, Charlie Gard, who had MDDS, Dominic Wilkinson and Julian Savulescu observe that treatment can be withdrawn on grounds related to wellbeing, noting that a life of severe disability, notably cognitive disability, counts against wellbeing. The authors add that, with the introduction of carrier testing for genetic conditions, parents need never have a child 'afflicted by' severe genetic disorder: testing 'could allow healthy children to be born instead of children who will at best have lives dominated by suffering and severe disability' (2018, p 165). Allen Buchanan et al claim that there are some conditions that are 'incompatible with a life worth living'. However, instead of relying on the argument that infants are simply replaceable they assert that since disability is a harm, 'common sense moral judgment' holds that a woman is 'morally wrong' to go ahead and have a child known to have, or at risk of, a disability instead of waiting to have a 'normal' child, and this includes cases where life is considered to be worthwhile (2000, p 247).

In other disorders such as Down syndrome, Marfan syndrome, and Noonan syndrome, some people are only mildly affected while others develop more serious symptoms. Some genetic disabilities such as Huntington's disease may be considered severe, but the first signs normally appear in adulthood. Perhaps a very severe genetic disability is one that results in severe physical and mental disabilities, where individuals are dependent on 24 hour care throughout their lives, and health complications may lead to a shorter life span. One such disability is Rett syndrome. Notably, the Nuffield Council observes that 'there are many complexities in the relationship between genomes and individuals' wellbeing, so even if genome editing could ensure the child did not have the genetic variant associated with a particular disorder, the "health" of that future person could not be guaranteed' (2018, p 3).

Even though worthwhileness of life is a subjective concept, the severity of a genetic disability is typically dependant on appraisals by health professionals, though people with such disabilities or their families and carers may have a different view (Boardman and Clark, 2022). The 'disability paradox' whereby people with serious disabilities consistently report that they have good and excellent quality of life when, to the external observer, their life is undesirable or burdensome is perhaps a striking example of the ableist paradigm (Bogart, 2022). The evaluation itself of the severity of a disability has serious implications not only for treatment decisions but also for the continuation of life.[1] With research indicating that just over 82% of US physicians believe that individuals with serious disabilities automatically have a worse quality of life than non-disabled people (Iezzoni et al, 2021), the stakes are high. This evaluation tends to take place in relation to disabled foetuses and born human beings who do not have capacity. Regarding foetuses, evaluation relies on prediction since the actual condition of the foetus may not be clear. This may also apply, to a lesser extent, to the born human being, especially when predicting future disabilities. However, in many cases, such as in Tay-Sachs, Angelman syndrome, and Rett syndrome, the disability only emerges after a few months of life after birth, hence the suggestion that infanticide has a place in healthcare (Savulescu, 2013). Indeed, infanticide is a legal option in some countries such as with the Groningen Protocol in the Netherlands. In addition, to ableist assumptions, evaluation may implicitly be affected by other factors notably notions that the very young are replaceable or have no personal status because theirs is not yet a lived personal life. Decisions to treat, or not to treat, are made with the expectation and possibly hope that the individual will die; and direct euthanasia is a logical extension of this (Singer and Kuhse, 1985). Of interest is what has been described as the 'better dead than disabled syndrome' whereby euthanasia is portrayed as an act of kindness or an application of beneficence in the case of severe disability and as a justification for declaring a life to be unworthy of life (Nelson, 1994).

If a life unworthy of life exists?

Inevitably, the concept of a 'life unworthy of life' raises the spectre of the 1920 pamphlet written by the German jurist, Karl Binding and the German doctor and medical professor Alfred Hoche. Titled *Allowing the Destruction of Life Unworthy of Life: Its Measure and Form,* Binding and Hoche presented the case for euthanasia in the form of the question 'are there any human lives, which have so strongly lost the quality of being the object of legal protection, that their continuation both for the person in question and the society has permanently lost any value?' (1920, p 14). Binding and Hoche's real concern was that these individuals were surviving into adulthood. Noting that someone might have become used to 'treasure the value of the single life for the individual and for the community' (1920, p 14), possibly due to the problematic influence of Christianity (1920, p 36), Binding drew attention to the wasteful loss of 'the most valuable life' in war or accidents in contrast to the 'burden' of care and expense spent on totally worthless life (1920, p 15). Hoche added the social burden of tying up healthcare personnel in unproductive tasks related to these 'empty human shells' (1920, p 35). This worthless life included 'weaklings', those who had suffered 'mental death', that is those with intellectual disability, and 'incurable idiots' whose lives were 'totally without purpose' but who did not see life as 'unbearable' (1920, p 18). Not only did the lives of these 'idiots' produce 'a terribly heavy load' for relatives and society, and so 'misuse' productive lives for an 'unworthy purpose', these people comprised 'the dreadful counterpart to real people and quickly arouse horror in nearly anyone (obviously not everyone) that encounters them' (1920, p 18). Consent or the presence of suffering were irrelevant. Binding and Hoche set out what they saw as the rational justification of euthanasia based on 'cool calculated logic', legal formulation, and scientific clarity (1920, pp 15–16). Their standing as members of the legal and medical professions gave weight to their proposition. While Binding acknowledged that their 'businesslike formulation' sounded 'heartless', he claimed that 'in truth, it stems only from the deepest compassion' (1920, p 14). In contrast, Hoche argued that it was a mistake to suggest compassion applied to the elimination of a worthless life, since where there was no suffering there could be no co-suffering (1920, pp 38–39). Looking forward to the future the authors concluded 'we may however one day mature to the view that the disposal of the fully mentally dead does not represent any violation, any immoral act, any instinctive brutality, but rather an allowable useful act' (1920, p 37).

Unworthy life or unworthy treatment?

What is to be done with a life considered to be unworthy of life is not a new issue: as the Roman Stoic philosopher Seneca (4 BC–AD 65) wrote, 'we put down mad dogs; we kill the wild untamed ox; we use the knife on sick sheep to stop their infecting the flock; we destroy abnormal offspring at birth; children too, if they are born weak or deformed, we drown. Yet this is not the work of anger but of reason – to separate the sound from the worthless' (c.45, 15.2). However, the belief that there is such a concept as a life unworthy of life is not simply a matter of direct killing: the belief also leeches into treatment decisions and decisions not to 'prolong' a life considered unworthy of life. Notably, this life may not be in the dying phase and includes decisions based on future quality of life. Significantly, some of these decisions explicitly accord certain human beings lesser moral status than other human beings. Put in contemporary and less prosaic terms than Seneca's, Wilkinson explains that 'infants appear to be benefited by continuing to live as long as that future life would contain more intrinsically positive than negative experiences (that is, they would have a life worth living)' (2013, p 131). He then argues that infants have a relatively weak interest in their future life since 'the story of their life has barely started' (p 142). Add in 'impairment' or disability and this 'increases the intrinsically negative features of future life'. Severe cognitive impairment in particular 'reduces the benefits of life' thus weakening further the interest in their future life (p 144). The benefit of life-sustaining treatment for a disabled infant may then compete against other stronger interests not to treat. As Wilkinson puts it, 'the harm to them if they die, is arguably less than that of an older individual' (p 152). Wilkinson claims that 'relevant differences' between the foetus and the newborn further weaken the former's interests (pp 149–150).

Unlike Seneca, Wilkinson situates his conclusions in beneficence. For Wilkinson, the birth of a severely disabled infant inevitably harms the family and, in many cases, the continued life of that infant is more harmful than the infant's death (2013, pp 117–119). As do Kuhse and Singer, Wilkinson details the negative experiences, consequences, burdens, and therefore considerable harm of a continued life to the family and child (2013, p 151). Since death for an infant is the absence of experience while continued existence with a life not worth living is a negative experience, he concludes that in many cases of severe disability, particularly the risk of future severe cognitive impairment, continued life is not in the child's best interests (2013, p 248). Moreover, he accepts without question that infants are replaceable (2013, p 108). For Wilkinson, a life worth living is related to wellbeing which includes the potential to have desires and preferences fulfilled. Notably for the purposes of disability, and perhaps in order to avoid a subjective account, Wilkinson adds that even if parents request continued treatment and even if from the child's perspective life is worthwhile, 'clear and convincing evidence' that the child will probably have severe impairment indicates that treatment should not be provided (2013, p 294). The 'probability test is less stringent than beyond reasonable doubt': where there is uncertainty in the future condition of the infant, Wilkinson argues that a 'keep-alive mistake' would be more serious than a 'let-die mistake' if the life turns out not to be worth living and so the threshold for providing treatment should be higher than where harm and benefit balance (2013, p 248).

Wilkinson relates his threshold principle to the worthwhileness of life rather than the more clinical decision with a longer provenance of the worthwhileness of treatment because he finds notions associated with treatment unsatisfactory. According to Wilkinson and Savulescu (2018), the concept of futile treatment has been heavily criticized on the grounds that it is both difficult to reach agreement on the concept and difficult to apply. Futility may relate to treatments that are of no effective benefit, unlikely to work, or they would be unwise for a doctor to prescribe. Part of the difficulty is the manner in which people interpret the goal of medicine: if the goal is to cure or to improve the patient's condition, then treatment that continues life or maintains the patient in their existing condition can be described as futile. Moreover, the authors claim that in most cases where there is a dispute about providing treatment, there is some chance of benefit. Therefore, they argue that what counts is if there is 'benefit enough' to provide treatment. They then bring in the notion of harm to either the child or to others as a reason justifying no treatment. There can then be a weighing up of benefits and burdens. However, they interpret this as the pros and cons not of treatment but of life versus death. Two significant factors count against prolonging life by continued treatment: reduced quantity of life and reduced quality of life. For Wilkinson and Savulescu the best interests test is interpreted as being about 'the value of that child's life for them' (sic). And this is based on 'how well a life is going'. The authors agree that a subjective account of wellbeing may be inadequate and that, in an objective account, it would be difficult to come to an agreement on which values to include. Nevertheless, the authors agree that serious disabilities, and especially severe and profound cognitive disability, can tip the scales against prolonging life. Notably, in their discussion of fairness and resource allocation they take the example of a child with 'normal intelligence' or a child who will have a 'normal life' and suggest that, depending on the level of abnormality, it would not be unfair discrimination to choose the 'normal' child over one with profound cognitive impairment.

Binding and Hoche's disagreement on the issue of compassion is interesting given the contemporary focus on justifying euthanasia on the grounds of avoiding harm, or acting with beneficence or that the individual is 'better off dead'. However, Binding and Hoche foreshadow the presumption underlying the contemporary view of those who hold to the existence of lives unworthy of life: that there is a separation, and not merely a distinction, between (1) being alive, biological life, or merely being human, and (2) having a life worthy of living, having a certain quality of life, and being a person.

Assumptions about persons and ableism

Ableism takes it as given that there are psychological and physiological norms sufficient to distinguish normality from abnormality. Rather than celebrate the diversity that is the human family where being limited is an aspect of our contingent human nature, ableism implies a hierarchy of abilities. A life of disability falls below a normal life, with profound cognitive disability falling even further. The assertion that some lives are not worth living buys into ableism and rejects the idea of the equal moral worth of all human beings. This hierarchy and the assertion about unworthy life often carry a significant and problematic assumption: that the individual whose life this is, is not a person. As Joel Reynolds puts it, 'the "true," "real," "natural" genome is the one that is perfect, that has no defects. That which is defective is not really, not truly, as it should be – not *really,* not *truly,* not *fully* human' (2022, p 117).

The concept of the person is both value-laden and contested. On the one hand, being a person gives some protection notably from being killed; on the other hand, there are as many definitions of the concept of 'person' as there are philosophers. One dominant bioethical definition centres on the actual capacity to exercise rationality. On this account, unborn individuals, infants, human beings with severe cognitive disabilities and some with severe genetic disabilities are not persons. Separating out biological and biographical personal life, James Rachels explains, biological life is 'merely being alive' whereas biographical life is 'having a life' (1986, Introduction). Those who exist in biological life may have a life story and history, but they may not have a biographical life that enables them to shape their own life. Rachels adds that in 'subnormal lives' the value of life is diminished. Indeed, if an entity lacks the mental and physical abilities to engage in simple human activities it does not have human life (1986, Chapter 4). According to one of the founding fathers of bioethics, Joseph Fletcher, human beings whose mental life is below an IQ of 20 are 'sub-personal beings', have no rights, and are 'outside the pale of human integrity' (1979, pp 19–20).

In his discussion on challenges to human equality, the philosopher Jeff McMahan, argues for the permissibility of killing 'marginal' beings, including 'animals, human foetuses and infants, anencephalic infants, severely retarded human beings, and human beings with acquired cognitive deficiencies of varying degrees of severity'. These individuals are not persons but rather beings that are 'very different in seemingly important ways from morally innocent, cognitively normal adult human beings' (2002, p 251). The sub-title of his book is *Problems at the Margins of Life.* As with Singer and Kuhse, people with cognitive disabilities are 'problems'. Once placed on the margins of personhood they are then placed on the margins of life itself. In his '"two-tiered" account of the morality of killing', killing of persons, that is individuals above a certain threshold characterized by psychological capacity, is governed by 'a requirement of respect for their inherent worth'; killing of those falling below the threshold relies solely on counting up interests in remaining alive. Only individuals above the threshold have 'equal worth' (2008, p 94). Notably McMahan thinks that respect for persons can allow for killing should the person's life cease to be worth living (2002, p 243). Certainly, McMahan recognizes there are powerful challenges to his two-tiered account, and he suggests that a three-tier account might accommodate those individuals, such as children, with the potential for rational life. Nevertheless, McMahan argues that 'those who are congenitally and radically impaired', that is those who have genetic defects that mean they have no potential for rationality, cannot satisfactorily be included except in a gradualist account of moral status (2008).

Disability as a harm

To overcome the problem of implied inequality and discrimination, some of those wishing to assert the existence of a life unworthy of life rely on the assumption that a life of disability inevitably involves harm. Harm makes life unworthy, even if individuals themselves do not experience suffering. As the British philosopher John Harris explains, the view that disability in itself is a harm need not imply that people with disabilities are not equal citizens with equal rights and protections. Harris defines disability as 'a condition that someone has a strong rational preference not to be in and one that is in some sense a harmed condition' (2001, p 384). Moreover, and returning to the question of moral status, Harris argues that it is not discriminatory to choose which 'sorts of people to bring into existence' or 'which embryos or foetuses to allow to become persons' because 'there is no one who suffers adversely from the choice'. Given that for Harris 'disabilities always involve a "harmed condition"', he claims that 'it is never wrong to prevent the births of people with disabilities and often right so to do' (2001, pp 386–387).

While Harris claims that his definition of disability is not related to normal species functioning (2001), Reynolds points out that ableism necessarily conflates disability with harm or suffering. Living 'outside ableist norms' is a lack of a 'natural good' and therefore a harm that causes suffering. Moreover, Reynolds says that ableism hyperinflates this conflation so that disability becomes not simply a 'local harm' of, for instance, mobility or cognition, but a 'global harm' that affects the whole of the individual's wellbeing (2022, p 105). The individual becomes the disability. Significantly, this is not to say that there is no harm or suffering in the lived experience of disability. Indeed, the actual condition of the person with disabilities can influence treatment decisions especially where treatment is burdensome or futile for this particular person. The concern is the claim that the very existence of a disability as an abnormality is automatically a harm simply because it is below the norm and this is unrelated to actual lived experience.

The notion that a life of disability is sub-personal or sub-human and the claim that disability is a harm often overlap. These two assertions are played out in the assumption that genetic disability can be eliminated by screening embryos and foetuses (Brecher, 2011), or by preventing the 'creation of persons with disabilities' (Brock, 2001). Although already 'created', affected embryos are not implanted; affected foetuses are terminated. In the case of Down syndrome, albeit not a severe genetic disability, pre-natal screening in Iceland has resulted in a near 100% termination rate, leading to the claim that Down syndrome has been 'basically eradicated' (Stefansson quoted in Quinones and Lajka, 2017). Factors such as personhood and harm are assumed: pre-birth there is no person; disability is a harm for the individual and for society. Since the individual concerned is not a person, it can be replaced by a healthy individual thus reducing harm (Singer, 2011, pp 164–165). But the claim to eradicate disability seems to disregard the fact that the embryo and foetus already have human lives since this claim is positioned as eliminating a disability not eliminating an individual: in a synecdochical way, the embryo or foetus is identified as the disability. Thus, it is argued that this elimination does not carry a negative evaluation of an existing person's life with that disability. However, this is to forget that personhood is a contested notion, that embryos and foetuses do in fact exist, they have their own unique identity, and in some cases, they exist with severe genetic disability. Life itself does not appear to be a significant reality, indeed life appears to have only instrumental value.

Instrumental value and the good of life

According to the adage *operari sequitur esse*, acts follow being: without being there would be no acting, no valuing and no becoming. Being alive is foundational to any values that may be added to that life, and worth cannot be prior because in order to know the values of life we have to exist. However, this adage still allows for the separation of biological and biographical life, as well as regarding life as having a purely instrumental value, which entails that life has value only as a means towards something else, in this case, towards a worthwhile life. A life without the features that make life worthwhile is designated a life unworthy of life. Putting aside for the moment the argument that the instrumental value of life discussed so far is based on arbitrary criteria and ableist discriminatory attitudes, the question arises, does life nevertheless have intrinsic value? According to the legal philosopher John Finnis, 'all human societies show a concern for the value of human life' (2011, p 83). He adds that one of the 'basic requirements of practical reasonableness' is 'no arbitrary preferences amongst persons' (p 453). Finnis claims that human life in all its aspects, including bodily life, is a basic human good (p 86). Following Finnis, and in response to McMahan, Patrick Lee argues that it is not merely *my* life that is intrinsically good, but the life of all rational beings where to be human is to belong to the kind of beings that have rational natures. A rational nature is what constitutes humans as human beings: all human beings have a rational nature. As I respect my own life, I should show respect and care for lives like mine, whether or not these individuals can exercise a level of rationality. Thus, Lee argues that all human beings have equal fundamental rights (2013). Lee's position is of course open to the charge of speciesism. However, from a philosophical point of view, speciesism itself is a problematic notion. After all, if speciesism is the undue privileging of one species over another, then to privilege, say, higher apes, over other non-ape animals is speciesist. This is recognized but also defended on the grounds of expediency by Singer, one of the architects of the Great Ape Project that seeks to include the whole ape species in the 'community of equals' (Cavalieri and Singer, 1993, pp 304–312), thus undermining the essence of the speciesism argument. Moreover, the views of some anti-speciesist philosophers are seemingly not only disablist but also speciesist because they use characteristics of so-called normal *human* persons as the benchmark to raise the status of some non-human animals (Matthews, 2020, p 282).

Framing the discussion of a life unworthy of life in terms of the instrumental and intrinsic value of life may be compelling. However, it risks remaining on the level of the general. The claim that there are basic goods, life being one of them, is not without critique and counter critique (Hittinger, 1988; May, 1988). Lee's argument depends on the acceptance of the notion of a universal rational nature. However, a life unworthy of life is not simply a general concept: it relates to actual individuals and lived experience. And it is a matter of justice.

Life worthy of life: a matter of reasoning or living?

The discussion so far indicates that specifying unworthy life is far from straightforward and is based on discriminatory assumptions. As with defining personhood, attempts to define worth rest on uncertainty and arbitrary criteria, interpretation varies between authors. Worthwhileness is itself ambiguous (Fumagalli, 2018). Even supporters of the notion that there is life unworthy of life point to a range of views, uncertainty about 'the quality of future life', and 'uncertainty about what level of life is worth living' (Brick et al, 2020).

Responding to some high-profile cases of the withdrawal of treatment for severely disabled children in the UK, Brick et al (2020) conducted a survey to assess public views not about worthwhileness of treatment or whether to treat, but specifically about the benefit of life in some hypothetical cases. They define a life not worth living as 'a life in which future burdens for the individual outweigh benefits. There is negative net future well-being'. Participants were asked the extent to which they agreed with three statements: 'life has no benefit for this infant', 'life is worse than death for this infant', and 'life has benefit for this infant – it would be best for this infant to remain alive'. From their findings the authors conclude that 'in the most severe cases, most people agree that life is not worth living for a profoundly disabled infant'. They confidently make this claim even though they concede that their sample size was small and predominantly white as well as female, that there were 'significant numbers of neutral responses', that 'the responses may not be generalisable', and that 'a short survey is unlikely to give enough time and information to reflect on whether a particular life is worth living'. Markedly, their initial questions were explicitly framed around life not worth living.

Against this background, there is 'fragmentation' in groups such as the Rett community over the effort and money put into developing risky technological innovations or seeking a 'cure' and consequently fewer resources available to improve the lives of people with Rett syndrome (Clarke and Sheikh, 2018). Rarely are people with disabilities or their families and carers asked what they want or need. Instead of changing discriminatory attitudes or practices, the drive is to change or to eliminate the individuals themselves or replace them with 'normal' individuals.

Of course, Brick's rather limited survey can be countered by clear evidence of the 'disability paradox', by the many stories that abound online about people who do live well with severe disabilities and by networks of disabled people opposing the legalized killing of disabled people. Academic research confirms that people who live with genetic conditions say that they have good health and a good quality of life. Suffering and a high quality of life can coexist without contradiction (Boardman and Clark, 2022). It seems then that there is a notable difference between lived experience and the way in which genetic conditions are defined, screened, and evaluated. Moreover, attitudes are shaped by systemic failures in implementing equality provisions (Equality and Human Rights Commission, 2017), by prejudice, and by a lack of understanding about disability (SCOPE, 2018) as well as unchallenged assumptions. However, while attitudes may be revealed through the language of *suffering with* disability rather than *living with* disability, or where death is seen as in someone's 'best interests' or where 'we' think that 'they' would rather not exist, professionals should have more appropriate attitudes to patients. The view that it is better not to be disabled should not become the rejection of the individual's very being.

Life unworthy of life: a matter of injustice

Having observed at the outset that the notion of a life unworthy of life is based on certain implicit unjust assumptions and discriminatory attitudes towards people with disabilities, it is time to make explicit my own assumptions. And these are, to begin with, that all human beings, as individuals with both material and immaterial aspects united in one being, have unique and personal lives from their very beginning. Distinguishing between the life of a human being and the life of a person need not imply a total separation: bodily life is part of the individual's personal reality. Indeed, the concept of the person was not originally tied to possessing certain mental capacities but rather was related to the uniqueness of each personal life (Matthews, 2020). Certainly, the condition of a person's life may vary. Nevertheless, the bearer of that life remains a living human being with full inherent human dignity. This is not to say that disability does not bring with it harm, disadvantage, lack of opportunities, or suffering. Suffering is not unimportant. However, suffering does not alter the fact that the living human being is a person. Nor is it to say that the harms associated with disability are solely due to disabling attitudes as the social model of disability would have it (see Shakespeare, 2010). However, disability is not just about harm because disability is an aspect of a whole life. The complexities of a whole life, whatever its condition, cannot be neatly defined or pigeon-holed. Nor can a human life be reduced to one apparently graspable element. As the French philosopher Emmanuel Mounier (1905–1950) explains, the whole human being in his or her 'total reality' cannot be grasped in the way that existences like trees, stones, or even other animals can be apprehended (1949, pp xvi–xvii) or valued. Human life is imbued with a 'spirit of wonder' (1948, p 17). Paraphrasing the French philosopher Gabriel Marcel (1889–1973), Mounier observes that the person is 'the reality whose contents *cannot be put into an inventory*' (1949, p 35). Moreover, and in contrast to McMahan, Singer, and Kuhse, for Marcel human beings are not 'problems' or puzzles to be solved. When we encounter another human being, we are in the 'presence of a mystery' (1930, p 93). Marcel's philosophy resonated with Mounier's own lived experience: Mounier's infant daughter Françoise, suffered a rare reaction to a small-pox vaccination and she lived the rest of her life in profound disability. This illustrates that severe disability cannot be eliminated simply by attention to genetics or preimplantation screening and diagnosis or to infanticide. Moreover, the designation of a life unworthy of life conflates the individual who lives with a disability with the disability itself. This conflation not only obscures the person, it also ignores the contribution that people with the profoundest of disabilities can make to their immediate life situation, their family, and to society.

In dominant bioethical discourses the designation of a life as unworthy of life, in cases of very severe genetic disability, is an invitation to consider intentionally bringing about the end of the life of the individual with that disability. This killing is described as a neutral act, or an act of beneficence, depending on the attitude to disability. It is not described as unjust killing despite the implicit ableist and disablist assumptions that accompany the decision. In terms of unjust killing the term murder may be more appropriate even if emotive. According to the British philosopher, Elizabeth Anscombe (1919–2001), murder 'has to do with equality in a deeper way than any equality of shares in material advantages'. Rather than having a right to life, for Anscombe 'everybody has or ought to have a right not to be murdered'. Anscombe believes that it is a mistake to define murder simply as 'killing that is wrong' since once a certain kind of killing, such as mercy killing, is justified, that justification seems to take killing out of the category of murder. She adds that another mistake is to think that murder is purely a matter of the law. An example here would be to say that since mercy killing is not legally murder, mercy killing is not murder. As Anscombe explains, 'it is clear that there is a moral notion of murder' since 'a peculiar gravity attaches to the idea'. A just person excludes certain kinds of actions, intrinsically unjust acts, from consideration. When considering why we may not kill innocent people, Anscombe points out that human life and human action necessarily involves justice, answerability, and innocence. To kill the innocent, even in the face of suffering, is an act of injustice and a failure to respect that person's 'mystical value' (2005, pp 253–260). An attitude of *pietas* towards human life engenders real concern over the deaths of migrants in boats, victims of COVID in care homes, and human beings with severe genetic disabilities. In end-of-life decision-making, this is where a distinction between clinical decisions based on the burden or futility of treatment for this particular patient are more discerning than decisions based on the arbitrary criteria surrounding the worthwhileness of life.

Conclusion

'The question of worthy life has always been a question of disability' (Reynolds, 2022, p 154). Although the lived experience of people with disabilities is significant, the debate between those who believe that the notion of a life unworthy of life does exist, and those who believe that no life is unworthy of life, cannot satisfactorily be answered by trading off experiences or pointing to instances when life has a certain quality, or describing benefits and harms, or outlining what makes life worthwhile. This form of debate risks simply becoming an exchange of competing opinions. Instead, an analysis of underlying assumptions and attitudes indicates that those who believe that there is a life unworthy of life in the case of severe disability make unjust assertions about what it is to be a human person, as well as about human dignity and equality. Suggesting that disability can be eradicated ignores the fact that disability does not exist without someone (a person) who lives with that disability, even if that someone is at the very beginning of life. The 'better off dead than disabled' world-view (1) fails to recognize that people have and will always be born in unexpected ways, (2) is unable to see a world within a caring system, and (3) fails to appreciate the contribution that people with profound disability can make. In Reynolds words, 'worlds worth living in are built by and through caring systems made for all sorts of people' (2022, p 159). Our question should not be whether a life unworthy of life can actually exist in the case of very severe genetic disability, but what kind of a world do we need to build to be worthy of this human life with such a disability.

Note

1 Currently, under English law, abortions may take place in the first 24 weeks of pregnancy under certain conditions specified in the *Abortion Act 1967*. Yet, if the foetus has a 'substantial risk' of being 'seriously handicapped', a termination may take place up to birth.

References

Anscombe, G.E.M. (2005) *Human Life, Action and Ethics: Essays by G.E.M. Anscombe*, Exeter: Imprint Academic.

Binding, A. and Hoche, [1920] (2015) *Allowing the Destruction of Life Unworthy of Life: Its Measure and Form*, trans. C. Modak, Policy Intersections Research Center: Suzeteo Enterprises.

Boardman, F.K. and Clark, C.C. (2022) 'What is a "serious" genetic condition? The perceptions of people living with genetic conditions', *European Journal of Human Genetics*, 30: 160–169.

Bogart, K. (2022) 'The disability paradox', Available from: www.psychologytoday.com/gb/blog/disability-is-diversity/202209/the-disability-paradox
Brecher, B. (2011) 'What is wrong with eliminating genetically based disability?', *Public Health Ethics*, 4(3): 218–225.
Brick, C., Kahane, G., Wilkinson, D., and Caviola, L. (2020) 'Worth living or worth dying? The views of the general public about allowing disabled children to die', *Journal of Medical Ethics*, 46: 7–15.
Brock, D. (2001) 'Two moral issues about disability', *American Journal of Bioethics* 1(3): 1–2.
Buchanan, A., Brock, D., Daniels, N., and Wikler, D. (2000) *From Chance to Choice*, Cambridge: Cambridge University Press.
Cavalieri, P. and Singer, P. (1993) 'The Great Ape Project – and beyond', in P. Cavalieri and P. Singer (eds), *The Great Ape Project*, New York: St Martin's Griffin, pp 304–312.
Clarke, A.J. and Sheikh, A.P.A. (2018) 'A perspective on "cure" for Rett syndrome', *Orphanet Journal of Rare Diseases*, 13(1): 44.
Equality and Human Rights Commission (2017) 'Being disabled in Britain: A journey less equal', Available from: www.equalityhumanrights.com/sites/default/files/being-disabled-in-britain.pdf
Finnis, J. (2011) *Natural Law and Natural Rights*, Oxford: Oxford University Press.
Fletcher, J. (1979) *Humanhood: Essays in Biomedical Ethics*, Buffalo: Prometheus Books.
Fumagalli, R. (2018) 'Eliminating "life worth living"', *Philosophical Studies*, 175(3): 769–792.
Harris, J. (2001) 'One principle and three fallacies of disability studies', *Journal of Medical Ethics*, 27: 383–387.
Hittinger, R. (1988) *Critique of the New Natural Law Theory*, Notre Dame: University of Notre Dame Press.
Iezzoni, L., Rao, S., Ressalam, J., Bolcic-Jankovic, D., Agaronnik, N., Donelan, K., Lagu, T., and Campbell, E. (2021) 'Physicians' perceptions of people with disability and their health care', *Health Affairs*, 40(2): 297–306.
Johnson, H.M. (2003) 'Unspeakable conversations', *NY Times Magazine*, 16 February, Available from: www.nytimes.com/2003/02/16/magazine/unspeakable-conversations.html
Lee, P. (2013) 'The basis for being a subject of rights: The natural law position', in J. Keown and R.P. George (eds), *Reason, Morality and Law: The Philosophy of John Finnis*, Oxford: Oxford University Press, pp 236–248.
Marcel, G. [1930] (2001) 'On the ontological mystery', in P. MacDonald (ed), *The Existentialist Reader: An Anthology of Key Texts*, New York: Routledge, pp 86–107.

Matthews, P. (2020) *Discerning Persons: Profound Disability, the Early Church Fathers, and the Concept of the Person in Bioethics*, Washington, DC: Franciscan University Press.

May, W. (1988) 'A critique of the new natural law theory by Russell Hittinger', *The Linacre Quarterly*, 55(4): 85–87.

McMahan, J. (2002) *The Ethics of Killing: Problems at the Margins of Life*, Oxford University Press.

McMahan, J. (2008) 'Challenges to human equality', *The Journal of Ethics*, 12(1): 81–104.

Mounier, E. (1948) *Existentialist Philosophies: An Introduction*, trans. E. Blow, London: Rockliff.

Mounier, E. [1949] (1952) *Personalism*, trans. P. Mairet, London: Routledge.

Nelson, J. (1994) 'Broken images: portrayals of those with disabilities in American Media', in J. Nelson (ed), *The Disabled, the Media and the Information Age*, Westport, CT: Greenwood Press, pp 1–24.

Nuffield Council on Bioethics (2018) 'Genome editing and human reproduction: social and ethical issues', Available from: www.nuffieldbioethics.org/assets/pdfs/Genome-editing-and-human-reproduction-short-guide.pdf

Perry, R. (2008) 'It's a wonderful life', *Cornell Law Review*, 93(2): 329–400.

Quinones, J. and Lajka, A. (2017) '"What kind of society do you want to live in?" Inside the country where Down syndrome is disappearing', Available from: www.cbsnews.com/news/down-syndrome-iceland/

Rachels, J. (1986) *The End of Life: Euthanasia and Morality*, Oxford: Oxford University Press.

Reynolds, J. (2022) *The Life Worth Living: Disability, Pain, and Morality*, Minneapolis: University of Minnesota Press.

Savulescu, J. (2013) 'Abortion, infanticide and allowing babies to die 40 years on', *Journal of Medical Ethics*, 39(5): 257–259.

SCOPE. Dixon, S., Smith, C., and Touchet, A. (2018) 'The disability perception gap: Policy report', Available from: www.scope.org.uk/campaigns/disability-perception-gap/#:~:text=The%20research%20shows%20that%20the,over%20the%20last%2020%20years.

Seneca (c.45) (1995) 'On anger', in J.M. Cooper and J.F. Procopé (eds), *Seneca Moral and Political Essays*, Cambridge: Cambridge University Press.

Shakespeare, T. (2010) 'The social model of disability', in L.J. Davis (ed), *The Disability Studies Reader*, New York: Routledge, pp 266–273.

Singer, P. and Kuhse, H. (1985) *Should the Baby Live? The Problem of Handicapped Infants*, Oxford: Oxford University Press.

Singer, P. (2011) *Practical Ethics*, Cambridge: Cambridge University Press.

UN (2021) 'Disability inclusive language guidelines', Available from: www.ungeneva.org/sites/default/files/2021-01/Disability-Inclusive-Language-Guidelines.pdf

Wilkinson, D. (2013) *Death or Disability? The 'Carmentis Machine' and Decision-Making for Critically Ill Children*, Oxford: Oxford University Press.
Wilkinson, D. and Savulescu, J. (2018) *Ethics, Conflict and Medical Treatment for Children. From Disagreement to Dissensus*, London: Elsevier.
Wong, A. (ed) (2020) *Disability Visibility: First Person Stories from the Twenty-first Century*, New York: Vintage Books.

7

Heritable Genome Editing and the Expressivist Argument

Calum MacKellar

Introduction

A lot has already been written over the years concerning the expressivist argument (sometimes called the 'expressivist objection'). This is used to argue that when actions, decisions, or indirect statements are made by a person, in certain circumstances, these may express undisclosed moral values, such as a discriminatory value system, just as clearly as if this had been revealed more directly through an oral or written declaration. Moreover, because these moral values have become transparent, this could have real consequences for others. Accordingly, the expressivist argument could include the following stages:

- An action, decision, or indirect statement is made;
- This action, decision, or indirect statement can only make sense, rationally, if the actor has a certain value system, which was not previously disclosed;
- This means that the actions, decision, or indirect statement may reveal a certain value system;
- This expressed value system of the actor may have consequences for others.

The expressivist argument reflecting a discriminatory value system may come in several versions[1] and may include a number of ethical arguments.[2] The Norwegian health scientist, Bjørn Hofmann, explains that 'one may very well argue that there is not one expressivist argument with many elements, but many (expressivist) arguments loosely related' (Hofmann, 2017, p 15). As a result, because the expressivist argument may be used in different ways, especially relating to reproductive procedures, it has sometimes been accused of being vague or even useless. Moreover, a number of counterarguments have been suggested.[3] This means that it may be helpful to unpack the different ways in which the expressivist argument has been considered in the context of existing selective reproductive procedures. This will then be used to examine whether the argument is relevant for selective heritable genome editing.

Background to the expressivist argument

The expressivist argument was developed by the American philosopher, Allen Buchanan (Buchanan, 1996), to reflect the notion that procedures seeking to correct, improve, or prevent, disability may assume and communicate a negative valuation of persons with disability (Boardman and Thomas, 2023). This was then used by some commentators, such as the American bioethicist Adrienne Asch (1946–2013), who was disabled herself, to suggest that the deselection of possible future persons with disability, using certain reproductive technologies, may reflect a discriminatory attitude towards persons with a disability.

In addition, it may send a negative message about the lives of these persons, affecting not only those who are already living with disabilities but also shaping societal attitudes toward them.[4] In this respect, the expressivist argument was presented in the following way by the British social scientists, Felicity Boardman and Gareth Thomas:

> The 'expressivist objection' refers to the notion that using reproductive (genetic) technologies to prevent the birth of future would-be disabled people contain, and express, a negative valuation of life with disability. (Boardman and Thomas, 2023, p 1223)[5]

This means that concerns relating to the implications of the expressivist argument may be present for any kind of screening and selective technology, such as prenatal screening, preimplantation genetic selection, and selective termination.[6]

In view of this, disability rights groups have often been the most vocal exponents of the expressivist argument in their critique of the discriminatory messages, attitudes, and ideas they believe are present in society.[7] It is also worth noting that the expressivist argument is relevant for any kind of selection based on biological traits including those that are considered positive in society. For example, if reproductive selection is ever made possible for good looks or being sportive, this may give a negative message to those who may not be considered to be physically beautiful or sportive.

But, as already mentioned, a number of counterarguments to the expressivist argument have been presented. For instance, the British sociologist, Sir Tom Shakespeare, who has a disability, indicated with respect to terminating pregnancies of disabled embryos/foetuses (Shakespeare, 2005, p 223):

> Some expressivists … argue that the use of prenatal diagnosis expresses the belief that life with an impairment is not worth living or that disabled people's lives are not valuable. But here I want to agree with those who have argued that individual decisions to terminate pregnancies affected by disability do not necessarily imply a moral judgement on the value of disabled people.
>
> Clearly, some individuals and policies make such judgement. But I believe that most prenatal decisions do not reflect beliefs that disabled people are less morally important, that disabled people do not have the right to be in the world, that disabled people are a financial burden on society, that diversity is not important, or that disabled people should not be supported and respected.
>
> I believe that the usual grounds for selective terminations are the beliefs that a particular impairment will cause suffering and/or restricted opportunities for the potential child, and/or that it will cause difficulties and restricted opportunities for the other siblings or the potential parents.

In this regard, Shakespeare is correct to emphasize that circumstances may exist in the selection between possible future persons where the expressivist argument may not apply. This would happen when a selective decision is not based on a discriminatory value system. For example, some parents may be prepared to welcome *any* child into existence (without preferences) but may decide to avoid having a child with certain characteristics because of other reasons, such as a lack of societal support in caring for such a child.[8] In other words, when parents select against a child with certain characteristics (such as disabilities), the reason may simply be a recognition, or belief, that they themselves lack the financial, physical, psychological, or social resources necessary to look after such a child. They may be recognizing their *own* limitations or that of society, rather than regarding disabled and non-disabled children as being different in value and worth.[9] This means that the expressivist argument revealing a discriminatory value system will only be relevant to reproductive selective decisions in the situation where prospective parents *could cope* with a disabled child but still select for a non-disabled child.[10] Arguably, the decision to select by these parents would then express a preference for a non-disabled child over a disabled one in their hopes and ambitions for a family. A preference which would also reflect an ableist worldview (as described by Pia Matthews in her chapter) since no other reason would exist for such a choice. This means that the decision of prospective parents may sometimes need to be unpacked or clarified.

Against this background, the expressivist argument has only just begun to be examined from the perspective of heritable genome editing and disability (MacKellar, 2021). Thus, the following sections will seek to examine the argument from a more general perspective while considering whether it is relevant to pre-conceptive selection.

Expression that the value and worth of the lives of persons with disability are inferior

Message given *that the value and worth of the lives of persons with disability are inferior*

The first way the expressivist argument may be used is to suggest that any decision by a person to select-out possible future individuals, solely because of a specific trait preference, is openly expressing this person's value system and revealing his or her views concerning the status of existing individuals who were born with such traits. In other words, it is suggested that selective reproductive technologies may express the view that these individuals' lives are unequal in value in comparison to others.[11]

This means that when prospective parents make an explicit decision to select between possible future children, solely on the basis of their biological trait preferences, they then make a public, factual, and very clear discriminatory decision. This is because the decision would be meaningless if all children were considered absolutely equal in value and in worth. Of course, while the decision may not be discriminating against existing children as such, since they just remain possible future children in the minds of the prospective parents, the explicit implementation of this decision to then select is very real and remains discriminatory. For example, if a decision is put before a person whether to accept a child with Down syndrome, the disability ethicists, Hans Reinders, in the Netherlands indicates:

> In any given case, the only reasonable answer to the question of why a disabled child should not be born is by reference to what one thinks about the lives of people actually living with the same disorder ... One may certainly refrain from making this judgement explicit, but the evaluation as such must be implicit in one's reasoning. This suggests that the lives of existing people with disabilities are at least indirectly implicated in strategies of prevention, because without such evaluations these strategies could not make sense. (Reinders, 2000, pp 8–9)

This version of the expressivist argument is crucial since it infers that any decision by a person to select-out possible future children *solely* because of disability, or any other biological trait, must logically be based on the preferences this person has for different kinds of children. And, by doing so, the person is betraying and revealing his or her views concerning those who already exist with disability or certain other traits.

A development of this argument was given by the American bioethicist, Scott Kim, with the example of a man with Parkinson's disease describing his fear of becoming similar to another patient he observes in a more advanced state of the disease. In so doing, he portrays his revulsion of what he considers the pathetic appearance of this patient while expressing his deep concern and disgust of the prospect of living like this person. Of course, it may seem that the man is only expressing his own private reaction without sharing any value-judgements concerning the patient with advanced Parkinson's, but Kim then explains:

> Although the writer may have intended to illustrate why he himself does not want to live like the man portrayed, that illustration works by way of evoking in us a reaction about the man portrayed ... The writer is judging the life before him as not worth living; [but] that judgment is logically prior to and serves to justify why he does not want that life. Judgments about whether a life is worth living, even if it is ostensibly only about one's own life, is not a private judgment. What this author shows the reader is his belief that any life such as the one he sees is not worth living. (Kim, 2019, p 135)

And though the decision to choose between possible future children is often made in the context of an individual's private and immaterial mind, if this decision is implemented by an act of selection, it then becomes a very real openly disclosed decision with tangible consequences including that only certain children should be brought into existence. Thus, any decision to select between possible future persons may be a real expression of the actual value system and preferences of the individual making the choice (based on the perceived desirability of the possible future persons).[12] Indeed, without such a reasoning the decision would not make sense. Moreover, if society enables such a choice, it would mean that the inherent and *equal* worth of all persons no longer exists.

Many persons with a disability are very much aware that non-disabled children are generally preferred by future parents (Fletcher, 2002, p 21). This may encourage individuals with a disability to see themselves as different from those who are not disabled, not only because of their different abilities but also on the very basis of their identity of who they are. This is not surprising since non-disabled individuals are similarly influenced in their identity by what they understand as their positive traits. For instance, good-looking persons may consider themselves as being better than the rest because of their physical beauty.

Message received *by persons with disability that the value and worth of their lives are inferior*

Research does indicate that some persons with a disability do, in fact, find the availability of selective reproductive technology distressing, devaluing, or offensive.[13] This means that when the deselection of certain possible future persons with disability is considered, some existing persons with the same disability may receive the message that they are, themselves, less able or even less valued than those who are non-disabled. This is because they may not be able to perform or accomplish activities considered to be normal in certain settings. Moreover, as disability rights campaigners argue, such a way of thinking may be seen as unacceptable since it may infer that the lives of persons with a disability are less valuable than those of others. The International League of Societies for persons with Mental Handicaps indicated in 1994:[14]

> Modern scientific practices such as selection by prevention cannot help but be informed by paradigms about disability and the human ideal. Certain commonly held assumptions put people with disabilities at risk in relation to genetic and reproductive technology. These include such questionable assumptions as people with disabilities suffer because of their disability; human value is judged according to intellectual or physical qualities of the individual; the human race can and should be improved; and we need only consider genetic factors to comprehend human difference. If the application of genetic research is driven by these sorts of spurious assumptions, the lives and rights of people with disabilities may be at risk.

In other words, persons who consider that some possible future lives are undesirable or inferior are, in a way, indirectly indicating that the value and worth of the lives of existing persons in similar or in worse situations are also inferior and this may be considered offensive by such persons.

But for others (including a number of individuals born with a disability), the possible indirect messages received from selective procedures may not be problematic or question the equality of value and worth of the individual's very existence. However, it would be difficult to see how the decision by a person to select between possible future children based *solely* on their biological trait preferences, can be anything other than discriminatory.[15]

Counter arguments concerning the expression that the value and worth of the lives of persons with disability are inferior

It is important to now unpack some of the counterarguments which have been suggested relating to the expressivist argument in the context of considering that the value and worth of the lives of persons with disability are inferior.

Expressed message may be unclear or misunderstood

The first criticism of this version of the expressivist argument indicates that questions may be asked about the intended message being sent and the possible message received, which should be distinguished (Botkin, 2000, p 305). In addition, it is suggested that it is unclear how any selection procedure can 'express' a certain message or, in fact, any message (Hofmann, 2017). For example, the UK-based bioethicist, Søren Holm, indicated that 'whereas an agent has control over what he or she intends to express, he or she does not have exclusive control over what an action or utterance actually expresses' (Holm, 2008, pp 24–25). It may also be argued that the reasons behind individual decisions relating to selective reproductive technologies are so varied that no single message may be communicated. But this may then mean that it becomes incumbent on the individual making the decision to explain and clarify why he or she is making the decision.

Moreover, it is suggested that neither individual decisions or societal expectations to deselect persons with certain traits necessarily express disrespectful messages (Nelson, 1998). And this is especially the case if they are made in private (Murphy, 2011). For instance, the British bioethicist, John Harris, argued that to suggest, other things being equal, that it would be preferable not to create a person who will suffer, unnecessarily, does not mean that persons with disabilities are somehow less valuable or lesser persons than others (Harris, 2000). However, this still begs the question on what basis, and from what moral worldview, the decision to select was taken if this is not explained since persons who experience suffering do not have lives which are less valuable than others if Article 1 of the UN Universal Declaration on Human Rights is accepted. This states that: 'All human beings are born free and equal in dignity and rights'.

Expressed message only concerns the condition and not the individual

One of the most common objections to the expressivist argument is that any negative messages that may be communicated by selective reproductive technology may not concern the disabled persons themselves but only the disorder. In short, it is suggested that it may be possible to disvalue a condition that gives rise to disability while still valuing persons who have the condition in the same way as any other person (Malek, 2010; Shakespeare, 2006). For example, the US National Academies of Sciences, Engineering, and Medicine indicated:

> [U]nconditional love for a disabled child once born and respect for all people who are born with or who develop disabilities are not incompatible with intervening to avert disease and disability prior to birth or conception. (National Academies of Sciences, Engineering, and Medicine, 2017, p 127)

But this overlooks the synecdoche argument which indicates that a disability does not exist on its own but is always embodied in a person. This means that the only way to avoid a certain congenital genetic disability is to avoid bringing *persons* into existence with the genetic disability.[16] For example, it is because of this variant of the expressivist argument that a prohibition of sex selection is supported by the World Health Organization seeking to address the problem of sex discrimination between persons (World Health Organization, 2011). In other words, because making a choice between the sex of possible future persons involves having preferences based on experiences with, or knowledge of, living persons with a certain sex, it is suggested that such a choice expresses and reveals a value system which does not recognize the sexual equality of all persons in society (Health Council of the Netherlands, 1995, p 37). Accordingly, one very real concern is that any support for such sex preferences in persons is contrary to the basic equality of persons of different sex and may reinforce discrimination between male and female individuals.

Consequently, it is when a trait such as a disability is accepted as being part of a person's very identity that the synecdoche argument, in the context of the expressivist argument, may become relevant. In other words, selective practices in this case, may not only be considered as expressing a negative valuation of the trait but also of persons with the trait. Moreover, it is important to note that even though persons with disability and their families may accept, reject, or rework this version of the expressivist argument (Boardman and Thomas, 2023), it is difficult to circumvent.

Opposition to selection may mean opposing all treatments

Another objection indicates that if the negative message expressed in the use of selective reproductive procedures is accepted, then this may also justify rejecting any treatment of disability or illness.[17] In other words, it is argued that if selective reproductive procedures instigate discrimination between those with or without a disorder, it would mean that any type of medical treatment would result in discrimination between those with or without a disorder – which is absurd. Thus, it is suggested that, just as the prevention of an illnesses, such as a flu infection, does not send a discriminatory message to sick people with the flu, so the prevention of a disability does not send a negative message to disabled people (Edwards, 2004).

In response, however, it is worth noting that disabled individuals may see their disability as being deeply interwoven with their very existence and an important part of who they are. This is especially true for those who were brought into existence with a biological disability.[18] Shakespeare explains that one of the strengths of the expressivist argument is that it 'captures the extent to which impairment is part of the identity of many disable people, particularly those with congenital conditions: whereas non-disabled commentators see impairment as a separate aspect (like having influenza), disabled people argue that it is an important aspect of who they are' (Shakespeare, 2006, p 89). In a way, a comparison could be made with the manner most people consider their birth-sex and how they are seen by others. This is because, for many persons, their embodied maleness or femaleness is a crucial part of their existence and who they are.[19]

Accordingly, a number of persons born with a congenital disability may realize that, had they not been born with the disability, then another person would have existed, which is reflected in the non-identity argument (Parfit, 1976; MacKellar, 2019). A conclusion which may be different for those who become affected by the same disorder later in life, since they already exist and have a place in society. In other words, by giving a person who already exists a treatment, one is not deciding to select between persons but deciding to change the very same person.

Negative evaluation may take place but not discrimination

It may also be possible to argue that selective reproductive procedures may give a message that the lives of persons with disabilities are negatively evaluated, but do not discriminate against them, because having a preference is different from prejudice and discrimination. In this regard, the Australian philosopher, Peter Singer, indicated:

> It is one thing to argue that people with disabilities who want to live their lives to the full should be given every possible assistance in doing so. It is another, and quite different thing, to argue that if we are in a position to choose, for our next child, whether that child shall begin life with or without disability, it is mere prejudice or bias that leads us to choose to have a child without a disability. (Singer, 1993, pp 53–54)

Similarly, Holm suggests that it is more likely that deselection decisions may express a negative attitude than a discriminatory attitude. But he then indicates:

> If having a particular disability is an essential part of my personal identity, part of what I am, the mere fact that I know or have reasonable reasons to believe … that others evaluate that disability negatively may affect my sense of identity and social standing negatively. And I may justifiably feel that way even if the negative evaluation does not lead to any actual discrimination. (Holm, 2008, pp 24–25)

Thus, it is very important to consider the way persons with congenital disabilities understand their identity. Indeed, they may still feel that their very existence may be negatively evaluated, based on a discriminatory value system, without experiencing direct discrimination, as such, meaning that the expressivist argument, in this case, remains valid.

Expression that persons with disability should not exist

Message given *that persons with disability should not exist*

If the lives of persons with disability are considered to be less desirable than the lives of others, it is then possible to argue, if a choice exists, that it is acceptable to ensure that certain possible future persons with a disability should not come into existence. And this can have repercussions on persons with disability who already exist. Hofmann indicates: 'According to the expressivist argument the choice to use biotechnologies to prevent the birth of individuals with specific disabilities is an expression of disvalue for existing people with this disability' (Hofmann, 2017, p 505).

In this regard, the possible negative message being expressed relating to the very existence of disabled persons may not only be given by individuals who use reproductive technology but by the health policies of a community which support selection against the existence of disabled persons.[20] For example, the American physician and ethics commentator, Leon Kass, discusses how future persons may be considered:

> A child with Down's syndrome or with hemophilia or with muscular dystrophy born at a time when most of his (potential) fellow sufferers were destroyed prenatally is liable to be looked upon by the community as one unfit to be alive, as a second-class (or even lower) human type. He may be seen as a person who need not have been, and who would not have been, if only someone had gotten to him in time. (Kass, 1976, quoted in Hofmann, 2017, p 508)

In addition, the philosopher, Tim Bayne in Australia states, in the context of court cases in which some individuals may seek damages for the simple fact of being born with a disability (wrongful life cases), that:

> One of the many worries that courts have expressed about wrongful life cases is that a positive verdict would commit the court to the view that the plaintiff not only ought not to have been brought into existence, but would now be better off dead. Although the worry as stated is groundless, there is something both true and important to it. … the kinds of conditions that prevent life from being worth starting are just those that prevent it from being worth continuing. (Bayne, 2015, p 56)

This means that the intentional deselection of possible future persons solely on the basis of their disability, or even the availability of such a selection, may be understood as an indirect rejection of individuals with a disability.

Message received *by persons with disability that they should not exist*

Making sure that certain persons with a certain disorder are not brought into existence, could also communicate a negative message that certain existing persons with a similar disorder should not have existed – a message which may be upsetting and even offensive for them to hear.[21] For instance, McMahan indicates:

> A disabled person might well be prompted to think: "These [preconception screening] programs are intended to enable people to avoid having the kind of child that I was, and to prevent the existence of people like me, because they judge that our existence does not repay the effort required for our care." (2005a, pp 150–151)

Furthermore, the UK's legislation which prohibits prospective parents specifically selecting for a deaf child may give the message that such deaf persons should not be born – a position that a number of persons in the deaf community may consider offensive.

But this kind of reasoning has also been contested. For instance, the English philosopher, Baroness Mary Warnock (1924–2019), indicated (while reducing *persons* with hereditary diseases to just the diseases themselves):

> The one thing I do deplore is the attitude of what I think of as the disability lobby who think it's wrong to try and eliminate hereditary diseases because it is suggesting that they are unworthy of life. That is an absolutely terrible argument. The disability lobby is very vocal on this subject and it does form a real obstacle on progress. (Warnock, 2018, np)

This all means that it is necessary to again examine the counterarguments relating to the suggested message received by persons with disability that they should not exist.

Counter arguments concerning the expression that persons with disability should not exist

The lack of empirical evidence of harm resulting from deselection of persons with disability

One counterargument to this expressivist argument is that little evidence exists that selection is harmful or promotes attitudes which are harmful to persons with disabilities.[22] However, some empirical studies do in fact demonstrate that selective reproduction procedures can, and do, result in tangible emotional harms in many persons with a disability (Boardman, 2014b; Boardman and Thomas, 2023).

Identity of the persons with disability

It is also possible to argue that a logical contradiction exists when some disabled persons object to selection procedures by indicating that the negative message received is that: 'I should not have been born'. This is because the 'I' in the statement cannot logically exist before birth.[23] It is further contended that many other events can take place that may result in a possible future person not coming into existence (Harris, 2001).

In response, however, it may be recognized that, even though the arguments are logically correct, it does not address the problem of the message being conveyed (1) about the moral value system of the person making the choice and (2) to existing persons affected by disabilities.

Expressed message not convincing when considering serious disorders

It has also been suggested that the expressivist argument reflecting a discriminatory value system may only be relevant when disorders are not significant. The 2018 Nuffield Council on Bioethics working group indicated:

> The expressivist objection has some traction where the consequences of the introduction of technology include negative effects on existing people with disability, and particularly where the disablement is mild and, to some extent, socially constructed. This is not the case, however, with the inherited genetic disorders that are the most likely targets for genome editing, which manifest in ways that significantly affect quality and length of life. (Nuffield Council on Bioethics, 2018, p 82)

However, this argument is based on the concept of quality of life and how much suffering or pleasure a person may experience during his or her lifetime. But this varies between every member of society and cannot influence, in any way, the inherent equality of value and worth of life, as stated by Article 1 of the UN Declaration of Human Rights, which is necessary for a civilized society to survive. As Matthews explains in her chapter, those who believe that there is such a thing as a life unworthy of life in the case of severe disability make unjust assertions about what it is to be a human person, as well as about the concept of equality.

Indicating that persons with disability should not exist does not express a negative attitude to existing persons

It is also possible to argue that seeking to prevent the existence of certain categories of persons need not express a negative attitude to existing individuals in the same category. This may happen even if the persons in the category might feel offended.

For instance, it may be suggested that children should not be conceived to unmarried teenage mothers and that it may be appropriate to try to avoid such circumstances, at least as a general public policy rule. In other words, that it is reasonable to try to prevent girls becoming pregnant at a young age by some non-controversial means (such as proper supervision, education, and appropriate aspirations). This raises the concern of whether measures seeking to prevent children being born from unmarried teenagers might inadvertently convey a negative message to the children who were actually conceived in these circumstances. In other words, it is suggested that, as with people conceived with an inherited disorder, the person conceived to an unmarried teenage mother has part of his or her identity tied to the specific circumstances of his or her conception. It means that it is possible that someone who was born in such a context may hear a debate about preventing teenage pregnancies and then ask 'are you indicating that I should not exist?'

Yet it may be reasonable to try to prevent girls becoming pregnant at a young age. And, if this is acknowledged, then it may be argued that it is acceptable to seek to prevent certain categories of person coming into existence. In other words, it is suggested that wishing to prevent the existence of such persons need not express a negative attitude to existing persons.

In response, however, it is worth noting that a number of unethical procedures exist which should not be used to bring children into existence, such as in early teenage pregnancies or even rape. But by indicating that bringing children into the world in this way may be unethical, does not say anything about the value and worth of the resulting child, as such. There is a moral difference between:

- Believing that certain persons should not exist because of the way they were brought into existence while believing that all persons who exist are equal in value and worth, and
- Believing that certain persons should not exist because they have less value and worth based on the belief that all persons who exist are unequal in value and worth.

Moreover, even the child generated through rape may recognize that he or she should not have come into existence in this way. But all agree that (in a civilized society) all are born equal and should equally be welcomed into the world. This means that when the very existence of some persons (and not the way they came into existence) is preferred over others, and the selection is made, then the expressivist argument remains valid.

Further challenges to the expressivist argument and the deselection of persons with disability

In the previous sections, the expressivist argument revealing a discriminatory value system has been shown to be valid with all counterarguments being insufficient to address the suggested concerns, especially if the value and worth of all human life is considered to be absolutely the same. But more general counterarguments have been proposed which now need to be examined.

Selection does not mean decrease acceptance

To begin with, the validity of the expressivist argument has been questioned because the deselection of certain possible future persons due to a particular biological trait does not seem to challenge the support being given to living persons with such traits (Durkin and Gottlieb, 2009). Public policy has generally sought to develop an inclusive society. Consequently, the fear that reducing the number of persons with congenital disabilities will necessarily decrease empathy, acceptance, or the integration of those who are affected by them in society, may be unfounded.[24] For example, the (US) National Academies of Sciences, Engineering, and Medicine indicated:

> [T]he decades that saw the explosion of prenatal diagnosis (accompanied by selective abortion of affected fetuses) and preimplantation diagnostics (accompanied by selective implantation of nonaffected embryos) are the same as those in which public attitudes toward disability became far more accepting.[25] It would appear that "encouraging attempts to reduce the incidence of a genetic disease is compatible with continuing respect for those born with the disease and providing support for their distinctive needs".[26] (2017, p 127)

However, it may still be argued that a clear inconsistency exists in ensuring that certain individuals are not brought into existence while encouraging the unconditional acceptance of all individuals, irrespective of their genetic characteristics (Parens and Asch, 2000b). As Reinders explains:

> Curiously enough we find ourselves confronted by two different worlds with respect to disability. One is the world of normalization, which depicts disabled people as people not unlike ourselves, people who

> have been wronged by their unnecessary exile from ordinary life and who, therefore, deserve our support. The other world is the world of prevention, which depicts living with a disability as a fate that can be worse than death and offers a rationale for justifying the practices of selective abortion and infanticide in the case of the severely disabled children ... The first message says, "Since you're here, we're going to care for you as best we can," but the second says, "But everyone would be better off if you were not here at all". (2000, pp 3–4)

Therefore, it may be contended that if a possible child with a disability is not a problem for society, and society is not a problem for the child, then all children should be welcomed into existence and supported, in an equal manner, regardless of their biological characteristics (Asch, 1999).

Proportionality is important but autonomy is paramount

Another counterargument suggests that the possible risks arising from selection should be balanced against the right to reproductive autonomy. For example, the British ethics commentator, Steven Edwards, indicates with respect to terminations: 'one may choose to have an abortion in the knowledge that this will offend prolife groups, but it does not follow that the offence caused to these groups outweighs one's right to make such a choice' (Edwards, 2004, p 419). Similarly, Shakespeare indicates: 'That termination of a pregnancy affected by a particular impairment may be upsetting to an individual with that impairment should not preclude others from exercising the right to this choice' (Shakespeare, 2005, p 223).

In other words, it is suggested that being able to decide between possible future children may be seen as an individual's right to reproductive autonomy and that such a decision has priority over any potential harm created by expressing negative attitudes towards, or even offending, certain people (Edwards, 2004). In this way, the argument indicates that while deselections may be harmful, it remains less harmful than alternatives (Hofmann, 2017).[27] Similarly, it may be contended that although deselections may indeed be the cause of reasonable distress and may be objectionable, any restrictions relating to them are unjustified in a liberal society. The harm is not severe enough to justify enforced coercive state interference.

But others are less categorical about the place given to autonomy. For example, some commentators indicated that the risks arising from certain selection decisions may mean that a case may exist for restricting selection against some disabilities (Gyngell and Douglas, 2018). Moreover, it is still difficult to justify certain choices when these reflect a belief in the unequal value and worth of certain lives in society, which may eventually undermine the very basis of civilized society.

Conclusion

Having examined the different concepts present in the expressivist argument as it relates to discriminatory selective reproduction and their counterarguments, it is possible to conclude that the argument is indeed valid if the decision to select is based solely on preferences despite the concept of equality of value and worth of all persons being accepted by society. This is because any decision to select between possible future persons, solely on the basis of their biological trait preferences, does indeed imply an inequality between the lives of possible future persons. In other words, those making a decision to select (when it is shared with others or made public) are, in fact, revealing (even betraying) their real views and core beliefs concerning those who already exist with certain biological traits such as disabilities. Views and beliefs that persons with disabilities are not absolutely equal to others in value and in worth, thus contradicting Article 1 of the UN Universal Declaration on Human Rights.

As a result, for many persons (whether disabled or not), making a decision that individuals with a disability should not exist, when made solely on the basis of their biological trait preferences, expresses a deeply discriminatory message that individuals with a similar disability who already exist should have been deselected. A message which may again contravene Article 7 of the UN's *Declaration of Human Rights* which indicates that: 'All are entitled to equal protection against any discrimination … and against any incitement to such discrimination'.

This also means that the 'social justice and solidarity' principle of the 2018 Report of the Nuffield Council on Human Bioethics' may develop into a permanent ban on heritable genome editing since it states that (Nuffield Council on Bioethics, 2018, p 87):

> The use of gametes or embryos that have been subject to genome editing procedures (or that are derived from cells that have been subject to such procedures) should be permitted only in circumstances in which it cannot reasonably be expected to produce or exacerbate social division or the unmitigated marginalisation or disadvantage of groups within society.

Thus, because heritable genome editing would, inevitably, undermine the concept of equality in society, if the procedure is considered solely on the basis of biological trait preferences, it will always be expected to 'produce or exacerbate social division or the unmitigated marginalisation or disadvantage of groups within society'. Moreover, if the concept of inclusivity is seen as a genuine priority in society, and prospective parents can consider, in an appropriate and positive manner, the possibility of having, and looking after, a disabled child, then heritable genome editing will no longer be meaningful in any circumstance. This is because all possible future children (with whatever biological trait they possess) would be considered as being equally worthy of life making selection meaningless.

Notes

1 For a review of expressivist arguments, see Buchanan, 1996; Buchanan et al, 2000; Hofmann, 2017.

2 See, for example, Wendell, 2013; Saxton, 2000; Parens and Asch, 2000a; De Grazia, 2012, pp 102–106; Boardman and Thomas, 2023.

3 For a discussion, see Boardman, 2014a; Nelson, 2000; Edwards, 2004; Malek, 2010; Shakespeare, 2013; Nuffield Council on Bioethics, 2018, p 84.

4 See, for example, Parens and Asch, 2000a; Nuffield Council on Bioethics, 2018, pp 83–84.

5 See also Nuffield Council on Bioethics, 2018, p 82: 'The "expressivist objection" holds that such interventions express (hitherto implicit) negative social attitudes towards people with certain forms of embodiment and may even compound such attitudes, or exacerbate a social environment that is hostile to people with disabilities more generally'.

6 See, for example, Nuffield Council on Bioethics, 2018, pp 83–84; Hofmann, 2017; Parens and Asch, 2000a; Boardman and Thomas, 2023.

7 For a discussion, see Parens and Asch, 2000b; Saxton, 2000; Kitcher, 1997.

8 For example, see Shakespeare, 1999, quoted in Scott, 2007, p 15.

9 For example, see Gavaghan, 2007; Robinson, 2023, Reed, 2020, p 540.

10 Of course, if individuals do not want to become parents this does not say anything, generally, about the value and worth of possible future children. Individuals may decide not to have children for many different reasons. See MacKellar and Bechtel, 2014, pp 46–50.

11 See, for example, Saxton, 2000; Wendell, 2013.

12 Of course, it may be possible for an individual to select persons who already exist, such as friends or a spouse, based on their personal characteristics, but this is different from making a real selective choice between possible future children because they are seen as unequal in their very value and worth.

13 See, for example, Boardman, 2014a; Barter et al, 2017, Nuffield Council on Bioethics, 2018, pp 83–84.

14 See The International League of Societies for persons with Mental Handicaps, 1994, p 11, quoted in Reinders, 2000, p 9.

15 Even if a future parent decides to select a child for a biological trait, such as a large head, he or she is expressing the belief that children with such traits are preferable to children without them (for whatever reason) meaning that a selection between possible future children is being undertaken (based on the non-identity argument) which is discriminatory.

16 Some commentators indicate that the reduction of a possible future person to just a specific trait is inappropriate since it disvalues the whole person. See Parens and Asch, 2000a; and Boardman and Thomas, 2023.

17 See, for example, McMahan, 2005b; McMahan, 2006; Harris, 2001; Buchanan et al, 2000; Nuffield Council on Bioethics, 2018, pp 83–84.

18 This also reflects, to some extent, the Origin Essentialism argument. For a discussion, see MacKellar, 2019.

In a similar manner, it is interesting that a large number (and it may even be a majority) of persons with a congenital disability consider their disorder as an integral part of their personal identity and who they are, see Bricher, 1999; Boardman and Hale, 2018.

19 Though some persons may seek to change their sex.

[20] For a discussion, see Hofmann, 2017; Holm, 2008; Press, 2000; Shakespeare, 1998.

[21] See, for example, Saxton, 2000; Parens and Asch, 2000a; Devandas Aguilar, 2019, paragraph 23.

[22] This is discussed in Hofmann, 2017.

[23] See, for example, Shakespeare, 2013, p 119 mentioned in Hofmann, 2017, p 508.

[24] For a discussion, see Boardman and Thomas, 2023.

[25] References in the original Hernandez, Keys, and Balcazar, 2004; Makas, 1988; Steinbach et al, 2016.

[26] References in the original Kitcher, 1997, p 85.

[27] Shakespeare also indicated that: '[W]hile there is evidence that some disabled people feel offended and discriminated against by prenatal diagnosis, there is no strong empirical evidence that material harms to disabled people results from selective abortion' Shakespeare, 2006, p 96, quoted in Scott, 2007, p 56.

References

Asch, A. (1999) 'Prenatal diagnosis and selective abortion: A challenge to practice and policy', *Am J Public Health*, 89(11): 1649–1657.

Barter, B., Hastings, R.P., Williams, R., et al (2017) 'Perceptions and discourses relating to genetic testing: interviews with people with Down syndrome', *Journal of Applied Research in Intellectual Disabilities*, 30(2): 395–406.

Bayne, T. (2015) 'In defence of genethical parity', in D. Archard and D. Benatar (eds), *Procreation & Parenthood*, Oxford: Oxford University Press, pp 31–56.

Boardman, F. (2014a) 'The expressivist objection to prenatal testing: The experiences of families living with genetic disease', *Social Science & Medicine*, 107: 18–25.

Boardman, F. (2014b) 'Experiential knowledge of disability, impairment and illness: The reproductive decisions of families genetically at risk,' *Health*, 18(5): 476–492.

Boardman, F. and Hale, R. (2018) 'How do genetically disabled adults view selective reproduction? Impairment, identity, and genetic screening', *Molecular Genetics & Genomic Medicine*, 6(6): 941–956.

Boardman, F. and Thomas, G. (2023) 'Expressivist objections to prenatal screening and testing: Perceptions of people living with disability', *Sociology of Health and Illness*, 45(6): 1223–1241.

Botkin, J. (2000) 'Line drawing: Developing professional standards for prenatal diagnosis services' in E. Parens and A. Asch (eds), *Prenatal Testing and Disability Rights*, Washington, DC: Georgetown University Press, pp 288–307.

Bricher, G. (1999) 'A voice for people with disabilities in the prenatal screening debate', *Nursing Inquiry*, 6: 65–67.

Buchanan, A. (1996) 'Choosing who will be disabled: Genetic intervention and the morality of inclusion', *Social Philosophy and Policy*, 13(2): 18–46.

Buchanan, A., Brock, D.W., Daniels, N., and Wikler, D. (2000) *From Chance to Choice: Genetics & Justice*, Cambridge: Cambridge University Press.

De Grazia, D. (2012) *Creation Ethics: Reproduction, Genetics, and Quality of Life*, Oxford: Oxford University Press.

Devandas Aguilar C. (2019) *UN Special Rapporteur on the rights of persons with disabilities, Rights of persons with disabilities*, 17 December 2019, A/HRC/43/41, paragraph 23.

Durkin, M.S. and Gottlieb, C. (2009) 'Prevention versus protection: Reconciling global public health and human rights perspectives on childhood disability', *Disability Health Journal*, 2(1): 7–8.

Edwards, S.D. (2004) 'Disability, identity and the "expressivist objection"', *Journal of Medical Ethics*, 30(4): 418–420.

Fletcher, A. (2002) 'Making it Better: Disability and Genetic Choice', in Institute of Ideas (E. Lee Comp.), *Designer Babies: Where Should We Draw the Line?*, London: Hodder & Stoughton, pp 15–28.

Gavaghan, C. (2007) *Defending the Genetic Supermarket: Law and Ethics of Selecting the Next Generation*, London: Routledge-Cavendish, pp 113–114.

Gyngell, C. and Douglas, T. (2018) 'Selecting against disability: The liberal eugenic challenge and the argument from cognitive diversity', *Journal of Applied Philosophy*, 35: 319–340.

Harris, J. (2000) 'Is there a coherent social conception of disability?' *Journal of Medical Ethics*, 26(2): 95–100.

Harris, J. (2001) 'One principle and three fallacies of disability studies', *Journal of Medical Ethics*, 27(6): 383–387.

Health Council of the Netherlands (1995) *Sex Selection for Non-medical Reasons*, The Hague: Health Council of the Netherlands.

Hernandez, B., Keys, C.B., and Balcazar, F.E. (2004) 'Disability rights: Attitudes of private and public sector representatives', *Journal of Rehabilitation* 70(1): 28–37.

Hofmann, B. (2017) '"You are inferior!' Revisiting the expressivist argument', *Bioethics*, 31(7): 505–514.

Holm, S. (2008) 'The expressivist objection to prenatal diagnosis: Can it be laid to rest?', *Journal of Medical Ethics*, 234(1): 24–25.

International League of Societies for persons with Mental Handicaps (1994) *Just Technology? From Principles to Practice in Bioethical Issues*, North York, ON: L'Institut Roeher.

Kass, L.R. (1976) 'Implications of prenatal diagnosis for the human right to life', in J.M. Humber and R.F. Almeder (eds) *Biomedical Ethics and the Law*, Boston, MA: Springer, pp 313–327.

Kim, S.Y.H. (2019) 'Lives not worth living in modern euthanasia regimes', *Journal of Policy and Practice in Intellectual Disabilities*, 16(2): 134–136.

Kitcher, P. (1997) *The Lives to Come: The Genetic Revolution and Human Possibilities*, New York: Simon & Schuster.

MacKellar, C. and Bechtel, C. (2014) *The Ethics of the New Eugenics*, New York: Berghahn Books.

MacKellar, C. (2019) 'Genome modifying reproductive procedures and their effects on numerical identity', *The New Bioethics*, 25(2): 121–136.

MacKellar, C. (2021) 'Why human germline genome editing is incompatible with equality in an inclusive society', *The New Bioethics*, 27(1): 19–29.

Makas, E. (1988) 'Positive attitudes toward disabled people: Disabled and nondisabled persons' perspectives', *Journal of Social Issues*, 44(1): 49–61.

Malek, J. (2010) 'Deciding against disability: does the use of reproductive genetic technologies express disvalue for people with disabilities?', *Journal of Medical Ethics*, 36(4): 217–221.

McMahan, J. (2005a) 'Preventing the existence of people with disabilities', in D. Wasserman, J. Bickenbach, and R. Wachbroit (eds), *Quality of Life and Human Difference*, Cambridge: Cambridge University Press, pp 142–171.

McMahan, J. (2005b) 'Causing disabled people to exist and causing people to be disabled', *Ethics*, 116(1): 77–99.

McMahan, J. (2006) 'Is prenatal genetic screening unjustly discriminatory?' *Virtual Mentor*, 8(1): 50–2.

Murphy, T. (2011) 'When choosing the traits of children is hurtful to others', *Journal of Medical Ethics*, 37(2): 105–108.

National Academies of Sciences, Engineering, and Medicine (2017) *Human Genome Editing: Science, Ethics, and Governance*. Washington, DC: The National Academies Press.

Nelson, J.L. (1998) 'The meaning of the act: Reflections on the expressive force of reproductive decision making and policies', *Kennedy Institute of Ethics Journal*, 8(2): 165–182.

Nelson, J.L. (2000) 'Prenatal diagnosis, personal identity, and disability', *Kennedy Institute of Ethics Journal*, 10(3): 213–228.

Nuffield Council on Bioethics (2018) *Genome Editing and Human Reproduction: Social and Ethical Issues*, London: Nuffield Council on Bioethics.

Parens, E. and Asch, A. (2000a) *Prenatal Testing and Disability Rights*, Washington, DC: Georgetown University Press.

Parens, E. and Asch, A. (2000b) 'The disability rights critique of prenatal testing: reflections and recommendations', in E. Parens and A. Asch (eds), *Prenatal Testing and Disability Rights*, Washington, DC: Georgetown University Press, pp 3–43.

Parfit, D. (1976) 'Rights, interests and possible people' in S. Gorovitz et al (eds), *Moral Problems in Medicine*, Englewood Cliffs: Prentice Hall, pp 369–375.

Press, N. (2000) 'Assessing the expressive character of prenatal testing: the choices made or the choices made available? in E. Parens and A. Asch (eds) *Prenatal Testing and Disability Rights*, Washington, DC: Georgetown University Press, pp 214–233.

Reed, P. (2020) 'Expressivism at the beginning and end of life', *Journal of Medical Ethics*, 46(8): 538–544.

Reinders, H.S. (2000) *The Future of the Disabled in Liberal Society: An Ethical Analysis,* Notre Dame, IN: University of Notre Dame Press.

Robinson, H. (2023) 'Prenatal testing, disability equality, and the limits of the law', *The New Bioethics*, 29(3): 202–215.

Saxton, M. (2000) 'The disability critique: Why members of the disability community oppose prenatal diagnosis and selective abortion' in E. Parens and A. Asch (eds), *Prenatal Testing and Disability Rights*, Washington, DC: Georgetown University Press, pp 147–164.

Scott, R. (2007) *Choosing Between Possible Lives*, Portland, OR: Hart Publishing (Bloomsbury).

Shakespeare, T. (1998) 'Choices and rights: Eugenics, genetics and disability equality', *Disability and Society*, 13(5): 665–681.

Shakespeare, T. (1999) '"Loosing the Plot" Medical and activist discourse of contemporary genetics and disability', *Sociology of Health & Illness*, 21(5): 669–688.

Shakespeare, T. (2005) 'The social context of individual choice', in D. Wasserman, J. Bickenbach, and R. Wachbroit (eds), *Quality of Life and Human Difference,* Cambridge: Cambridge University Press, pp 217–236.

Shakespeare, T. (2006) *Disability Rights and Wrongs*, London: Routledge.

Shakespeare, T. (2013) *Disability Rights and Wrongs Revisited*, New York: Routledge.

Singer, P. (1993) *Practical Ethics*, 2nd edn, Cambridge: Cambridge University Press.

Steinbach, R.J., Allyse, M., Michie, M., Liu, E.Y., and Cho M.K. (2016) '"This lifetime commitment": Public conceptions of disability and noninvasive prenatal genetic screening', *American Journal of Medical Genetics*, 170(2): 363–374.

Warnock, M. quoted in Bhattacharya, S. 'Baroness Mary Warnock interview: on the 14-day rule, philosophers as committee chairs, and the next report', *BioNews*, 30 April 2018, No. 947.

Wendell, S. (2013) *The Rejected Body: Feminist Philosophical Reflections on Disability*, New York: Routledge.

World Health Organization (2011) 'Preventing gender-biased sex selection', Geneva, World Health Organization, Available from: http://whqlib doc.who.int/publications/2011/9789241501460_eng.pdf [Accessed 24 February 2020].

8

The Dark Side of CRISPR: Its Potential Ability to 'Fix' People at the Genetic Level is a Threat to Those Who Are Judged by Society to Be Biologically Inferior

Sandy Sufian and Rosemarie Garland-Thomson

Americans have celebrated the fact that the Biden administration is embracing science and returning the country to evidence-based policy making. We agree that science should guide policy – except in cases where it would not assist people to live their lives but would, instead, exclude them.

The CRISPR-Cas9 gene-editing technology, for which biochemists Jennifer Doudna and Emmanuelle Charpentier won the Nobel Prize in Chemistry, has the potential to do just that. So do other forms of scientific technologies. We should therefore always be aware of the ethical choices these technologies can pose.

In the case of CRISPR, those choices are complex. CRISPR has many functions; one of these is that it can be used to treat disease. Yet the far-reaching, more fraught promise of this technology – one about which scientists seem at once excited and cautious – lies in its ability to eliminate from the gene pool what medical science identifies as faulty or abnormal genes that cause difference in individual people. Certainly, goes the logic of CRISPR's promise, the goal of ridding future generations of terrible diseases that cause suffering and death and deplete resources, seems an unquestionable enterprise.

But Doudna herself has recognized that CRISPR carries with it 'great risk'. In a *New York Times* interview on 22 October 2020, she warned of the unknown consequences of embryo editing, cautioning researchers to wait to use CRISPR for these ends.

As disability studies scholars and women with genetic differences who are experts in thinking about the consequences this technology will have for actual human beings, we have grave worries that the use of these 'genetic scissors' will, in the future, cut people like us out of existence without others even noticing. Scientists who use CRISPR could see editing genes such as ours out of the gene pool as entirely uncontroversial.

This attitude, in fact, would be consistent with wider societal views. The idea that ridding society of genetic differences that count as disease or defect is an undeniable 'good' continues to be pervasive in our society. Americans generally see no problem with editing genes linked to broad swaths of people like us; after all, supporters of this view may argue, editing out a gene-linked condition is different from editing out a person, and curing disease is an indisputably good thing.

But our genetic conditions are not simply entities that can be clipped away from us as if they were some kind of a misspelled word or an awkward sentence in a document. We are whole beings, with our genetic conditions forming a fundamental part of who we are. Still, many Americans – including medical providers and even some people with genetic differences – consider lives such as ours as *not worth living as they are.*

Further, the common belief that ridding disease and anomaly from society is an incontrovertible good can lead very quickly from the actual possibilities of science to fantasies of 'improving' humanity where we would all become some aspirational version of personhood that is somehow better, stronger, smarter, and healthier. But CRISPR's tantalizing offer to achieve the supposedly 'best' kind of people at the genetic level is an uneasy alert to those who are often judged to be biologically inferior – one we know all too well. People like us whose being is inseparable from our genetic condition would be the first to go.

We both have genetic conditions that many people consider serious enough to eliminate from the human gene pool: one of us lives with cystic fibrosis (CF), and the other a form of syndactyly. Both of these conditions have shaped our bodies and our lives. Sandy's affected lungs require several hours of treatment each day, and Rosemarie's affected hands limit her manual dexterity. We are among the one billion people in the world (15 percent of the population) and 61 million people in the United States (26 percent of all adults) who are considered disabled. We are among the 10 percent of all adults who have a genetic condition.

Because we were born with our conditions, we have benefited from learning early on how to live with the characteristics of our particular genetic distinctiveness. Our supportive families saw to it that we accessed good health care and received educations suitable to our talents and interests. Improved medical treatments, social progress, and political equality movements raised our quality of life in ways that people like us in generations prior to ours could not have imagined.

When Sandy was born in 1967, people with CF had an average life expectancy of 15, but during 1970–1990, life expectancy doubled due to new medical therapies. Today's average life expectancy is 44, but with novel medicines called cystic fibrosis transmembrane conductance regulator (CFTR) modulators, people with CF are expected to live even longer with fewer hospitalizations. These transformations in life expectancy attest to the changing nature of prognosis, one for which CRISPR's editing cannot account.

When Rosemarie was born, in the late 1940s, people with physical disabilities like hers were often institutionalized and led limited lives far from the support of their families. At that time, only one in five children with disabilities were educated in public schools with non-disabled children. Physically disabled children were most often sent to segregated schools where they received inferior education. With the Education for All Act of 1975 (now the Individuals with Disabilities Education Act (IDEA)), however, the federal government guaranteed public education and services for all children with disabilities, thus changing their life trajectories.

We learned to thrive with the bodies we have and possess identities and lives that include our genetic diagnostic categories but also go beyond them. Yet stubborn beliefs about 'good' genes and 'bad' genes nonetheless persist in discriminatory attitudes that affect us both. When Rosemarie was pregnant with her first child, the obstetrician assumed that her major concern was that the baby would have hands and arms like its mother, even though Rosemarie's biggest concern was finding a good childcare situation that would complement her job responsibilities.

When Sandy considered having a biological child, friends and medical providers questioned her decision to consider pregnancy because that meant passing on one copy of her cystic fibrosis gene to a future child. This imagined child would not have had the disease since her husband is not a carrier for CF (CF is an autosomal recessive disease). But some of those in Sandy's circle still believed pregnancy inadvisable because, to them, producing a child who would carry the CF gene was equally undesirable. Sandy called out their assumption: that her condition was inherently inferior – a point they took as self-evident despite the fact that an estimated 24 percent of people worldwide are carriers for genetic conditions.

These stories also reveal an enduring ideology about the inextricable, cultural link among disability, reproduction, and suffering. They illustrate the subtle, yet insidious, idea that some genes are inherently bad and contaminate the human gene pool; as such, people who carry them should not propagate and pass those genes on to their progeny so as to make those children either carriers or affected. These ideas also expose an even deeper, ableist assumption: that people with supposedly 'bad genes' fundamentally suffer and hold a less valuable place in society than others.

This isn't to say that people with genetic conditions don't suffer, but we don't necessarily suffer *all the time* and we don't necessarily suffer any more than other people without such conditions. Yet the cultural impulse to assume that people with genetic variations are in a constant state of suffering, and that it blights our lives, is so pervasive that it is even internalized by some with genetic conditions themselves.

Such genetic determinism is a new form of eugenic thinking grounded in what the communications studies scholar James L. Cherney calls 'common sense' ableism, a belief system that allows people to simultaneously deny any commitment to distasteful eugenic principles while also holding them up. Common sense ableism permits, even encourages, such injurious attitudes.

Utilizing genome manipulation tools and performing genetic selection is tantamount to engaging in what Rosemarie calls 'velvet eugenics'. Enforced by laissez-faire commercialism, rather than by the state, velvet eugenics seems like common sense, yet it hides its violence and inequality behind claims of patient autonomy and under a veil of voluntary consent. Ultimately, market-driven velvet eugenics embodies a similar goal of purging unacceptable human variations that campaigns to eliminate the supposedly unfit and inferior have held in the past. Both enact a mandate to exclude people with disabilities from coming into the world.

People like us should not be edited out of existence in some version of a utopian future. This vision of a future without people like us limits our ability to live in the present. Evaluating the quality of life of another person is a complex, highly subjective, and context-dependent task that is morally questionable in a society based on the concept that all people are of equal value regardless of their individual differences. The limitations of human imagination make it questionable, if not unethical, for a person to grasp another person's (or group of people's) quality of life fully.

Expanding diversity in all its forms, including disability, strengthens the human community ethically and biologically because it opens the public and private sphere to a variety of perspectives, life experiences, ideas, and solutions to live together with mutual flourishing. More importantly, our shared founding belief in the equal value of all members of a society should remind us that people's worth should not be determined by social judgments about their contribution. All members of a community contribute to its welfare by existing in their individual distinctiveness.

Genome editing is a powerful, scientific technology that can reshape medical treatments and people's lives, but it can also harmfully reduce human diversity and increase social inequality by editing out the kinds of people that medical science, and the society it has shaped, categorize as diseased or genetically contaminate – people like us who are understood as having bad genes. But we should be reminded that bad genes do not necessarily lead to bad lives, just as good genes do not necessarily lead to good lives. If CRISPR is put to use to eliminate rather than to treat genetic difference, we as a society would essentially instrumentalize this moralistic and reductionist assumption.

Acknowledgement

This article was first published by *Scientific American* on 16 February 2021 and is reproduced with permission.

References

Balfour-Lynn, I.M. and King, J.A. (2022) 'CFTR modulator therapies – effect on life expectancy in people with cystic fibrosis', *Paediatric Respiratory Reviews*, 42: 3–8.

Cherney, J. (2011) 'The rhetoric of ableism', *Disability Studies Quarterly*, 31(3).

Lazarin, G.A., et al (2013) 'An empirical estimate of carrier frequencies for 400+ causal Mendelian variants: Results from an ethnically diverse clinical sample of 23,453 individuals', *Genetics in Medicine*, 15(3): 178–186.

Swisher, K. 'Podcast interview with Jennifer Doudna', *New York Times*, 22 October 2020.

Zhang, S. 'The last children of Down Syndrome', *The Atlantic*, December 2020.

9

Could Heritable Genome Editing Be Used Without Undermining Equality?

Richard Playford

Introduction

In the 1997 science fiction film *Gattaca*, directed and written by the New Zealander, Andrew Niccol, eugenics has become commonplace in the not-too-distant-future. As a result, parents may choose, prior to conception and birth, to put their child through a process of genetic screening and modification to suit their wishes and ambitions. Such children are subsequently classified as 'valids'. On the other hand, children born naturally, and who are thus susceptible to genetic disorders, are classified as 'invalids'. While genetic discrimination is illegal, in reality genetic profiling is commonplace resulting in widespread discrimination. Unsurprisingly, it is the valids who are the beneficiaries of this discrimination, particularly in employment, with invalids being relegated to menial jobs. The film then follows the struggles of an invalid to overcome this discrimination and to realize his dreams of becoming an astronaut.

Though many themes are explored, including issues of genetic determinism, parental expectations, and sibling rivalry, one of the major themes is that of discrimination and equality. While the film illustrates a scenario in which genomic engineering clearly does undermine equality while promoting discrimination, this naturally raises the question: is there a scenario in which it might not? Are there circumstances in which the benefits of genomic engineering can be enjoyed without any resultant inequality or discrimination? In order to answer this question, I will try to identify a substantive conception of equality which has relevance for this topic, before applying it to this question in order to determine possible recommendations.

Clarifications and assumptions

Before proceeding further, it is worth clarifying certain assumptions I will make in the course of trying to answer this question. I will assume that any changes we could make to our genome (and thus phenome)[1] will start off being small, perhaps comparatively simple interventions to prevent or treat severe genetic disorders, but that as the technology progresses, and our attitudes towards its use change, the potential modifications we could make will become more significant. I will assume that eventually significant changes could be made to both the human genome and, perhaps more importantly, the human phenotype, that is, our external and observable physical, physiological, and psychological traits. I will also assume that, eventually, these changes will be sufficiently significant so as to have a meaningful impact on our personal relationships, our power dynamics, and the (comparative) quality of our lives. Finally, I will assume that, for better or worse, these technologies are coming, and that people are going to use them. It should be noted that these assumptions are not entirely arbitrary but instead are based on the observation that it is common for once controversial medical practices to become permitted and then commonplace over time. Examples include research with embryos, preimplantation genetic diagnosis, as well as assisted suicide and euthanasia. All of these have either become, or are in the process of becoming, more and more common, at least in the countries in which they are permitted, despite the fact that, until relatively recently, they were highly controversial and, in some cases, almost universally banned.[2] As a result, we are looking for a theory that will give us some concrete guidance on how to regulate these technologies so as to maximize their benefit whilst minimizing their harms. Thus, the question takes on a practical and political hue and should be answered with these goals in mind. As we shall see, in answering this question I will lean on the work of two prominent contemporary philosophers: Peter Singer and John Rawls.

Equality of what?

As we have seen in previous chapters, there are many different conceptions of equality. Many of these have, at least at first glance, little to say on this topic, such as the distinction between numerical and qualitative identity.[3] Others, at least if considered in isolation and simplistically, are simply unaffected by the possibility of genetic engineering, for example, equality before the law is, *per se*, perfectly compatible with genetic engineering just so long as any laws concerning genetic engineering are applied to everyone equally. Others, such as equality of outcome, are clearly affected, but it is very unclear what we should do with this insight. A full discussion of all of these conceptions of equality goes beyond the scope of this chapter, and large swathes of such a discussion would risk being trivial or uninteresting as just illustrated. As a result, we need a non-trivial conception of equality that will say something useful about genetic engineering. It is worth noting that there is nothing to stop there being more than one such conception of equality. However, again, due to limits on space I cannot discuss all of the possible conceptions here. Instead, I will try to identify one such conception so as to advance our understanding of these issues and to provide at least some modest recommendations for how we might proceed.

Singer has proposed one such conception of equality. According to Singer's principle of equal consideration of interests we should 'give equal weight in our moral deliberations to the like interest of all those affected by our actions' (Singer, 2011, p 20).[4] Singer asserts this as a general moral principle which should govern us at both an individual and societal level and many scholars have raised serious objections to it as such (for example, see Zoulo, 2017). Additionally, as previously explained, our interests are more political in nature than those of Singer when formulating his principle. By this I mean that our interests are not in how we as *individuals* (in an ethical context) ought to deliberate about equality, rather our interests concern *society as a whole* and how we, as a political community, *ought to structure and govern society* (politically) when considering issues of equality. As a result, I suggest we therefore modify Singer's principle and adopt the following principle:

> The political principle of equal consideration of interests: when formulating legislation and enacting policy we should give equal weight to the interests of all members of our community.

Given this principle is only applicable in a political context, and given it is perfectly compatible with various other political (and ethical) principles, I suggest this principle is less contentious than Singer's more general ethical formulation. Further, in a political context, the principle seems intuitively plausible and likely to offer sensible suggestions when formulating public policy. There are many different members of our society, all with differing and potentially competing interests. It seems to follow from this, almost of logical necessity, that if we are to treat them equally, we must weigh up those interests equally.

Some may object to this principle on the grounds that there are times when we do not want to treat everyone's interests equally. For example, if the state needs to raise taxes to fund a large infrastructure project, we might believe that everyone has an equal interest in not paying these taxes. We might then conclude that, if we are to follow this principle, taxes should be distributed equally among all members of society through a poll tax, with the very wealthy and the very poor shouldering the burden 'equally'. Many would conclude that such a policy would be unjust because it would hit the poorest in society hardest and wealthiest in society least. Therefore, some might conclude that such a principle is implausible. Relatedly, some might think that such a principle risks trampling on the rights of minorities. If the majority have an interest, for example, in oppressing a minority in some way then perhaps, even though the minority have an interest in not being oppressed, we might think that this principle simply recommends weighing up those competing interests 'equally' and thus favouring the majority at the expense of the minority in an oppressive and unjust way.

However, this misunderstands the principle. As Singer points out when discussing his principle, albeit in an ethical context:

> Equal consideration of interests is a minimal principle of equality in the sense that it does not dictate equal treatment. Take a relatively straightforward example of an interest, the interest in relief of physical pain. Imagine that after an earthquake I come across two victims: one with a crushed leg, in agony, and one with a gashed thigh, in slight pain. I have only two shots of morphine left. Equal treatment would suggest that I give one to each injured person, but one shot would not do much to relieve the pain of the person with the crushed leg. She would still be in much more pain that the other victim, and even after I have given her one shot, giving her the second shot would achieve a more marked reduction in her pain than giving one shot to the person in slight pain would do for that person. Hence equal consideration of interests in this situation leads to what some may consider an inegalitarian result: two shots of morphine for one person and none for the other. (Singer, 2011, p 22)

Likewise, our political principle of equal consideration of interests would potentially allow us to implement a progressive tax rate. Whether it actually would be equal (or fair) in the real world is obviously a more complicated question, but our principle would certainly not rule it out and would even endorse it in certain situations. As Singer himself points out '$100 means a lot to someone for whom it is equivalent to his weekly income, but it means very little to a billionaire' (Singer, 2011, p 23).[5]

These same sorts of considerations can be used to respond to the concern that our principle might allow for the persecution of a minority if doing so is in the interests of the majority. It is not uncontentious to say that under most, if not all, realistic circumstances a minority has a much stronger interest in not being persecuted than the majority does in persecuting them for some, presumably far less significant, putative benefit.

Additionally, as previously stated, our political principle does not rule out the adoption of other principles, alongside our political principle of equal consideration of interests, when formulating public policy. Many such principles are likely to protect the rights of minorities, any number of which could be adopted. For example, an account of human rights seems likely to do this. It is worth noting that this same strategy is not available to Singer because his ethical principle of equal consideration of interests is a more general ethical principle (which forms the backbone of his broader consequentialist utilitarian ethical theory). This may give us reason to doubt Singer's ethical theory, but it need not concern us when considering our political principle of equal consideration of interests. As a result, I propose we adopt this principle going forward.

Whose interests?

We now have a principle with which we can tackle our question. However, as it stands, our principle is still ambiguous. It states that we should give equal weight to the interests of all members of our (human?) *community*. As a result, the question now becomes, who counts as a member of our community?

In order to answer this question, I suggest we adopt Rawls's account of moral personality. To claim that someone has moral personality is not to claim that they are necessarily a good person, rather it is to claim that they have the two key features that make someone a person from a moral perspective. According to Rawls moral persons are distinguished by two features:

1. 'They are capable of having (and are assumed to have) a conception of their good (as expressed by a rational plan of life); and …'
2. 'They are capable of having (and are assumed to acquire) a sense of justice, a normally effective desire to apply and to act upon the principles of justice, at least to a certain minimum degree'. (Rawls, 1999, p 442)

In a sense, all Rawls is doing here is offering an account of personhood, but with a particular emphasis on the moral aspects of being a person. Such an account allows us to clearly articulate whose interests we should consider when following the political principle of equal consideration of interests. Singer himself objects to Rawls's account of moral personality on the grounds that:

> [N]ot all humans are moral persons, even in the most minimal sense. Infants and small children, along with humans with profound intellectual disabilities, lack the required sense of justice. Shall we then say that all humans are equal, except for very young or intellectually disabled ones? This is certainly not what we ordinarily understand by the principles of equality. (Singer, 2011, p 18)

Singer then considers a response Rawls could make. He writes: 'Rawls deals with infants and children by including potential moral persons along with actual ones within the scope of the principle of equality' (Singer, 2011, p 18). He then responds to this by saying that:

> This is an ad hoc device, confessedly designed to square his theory with our ordinary moral intuitions, rather than something for which independent arguments can be produced. Moreover, although Rawls admits that those with irreparable intellectual disabilities 'may present a difficulty', he offers no suggestions towards the solution of this difficulty. (Singer, 2011, pp 18–19)

Unfortunately, Singer's criticism misrepresents Rawls's position, and thus it fails as a criticism (both of Rawls and of my argument here).[6] Rawls does not believe that young infants are merely potential moral persons. Rather, he defines moral personhood in such a way so as to include very young infants. As Rawls explains: 'One should observe that moral personality is here defined as a potentiality that is ordinarily realized in due course. It is this potentiality which brings the claims of justice into play' (Rawls, 1999, p 442). Rawls's point is simply that it is not the ability to actually think about issues of morality or to conceptualize one's own good in the here and now that makes one a moral person. Rather simply being the sort of thing that, when mature, healthy, educated, and so on, usually can do those sorts of things is sufficient for moral personality.

We can also now identify a strategy we can use to justify the claim that those with severe intellectual disabilities also count as moral persons even though they lack, and will likely always lack, the ability to have a conception of their own good and a sense of justice as explained earlier. We need not abandon the claim that such individuals are moral persons because while they cannot *actually* do these things in the here and now, there is a sense in which they still have the *potential* to do these things. Sadly, however, this potential is blocked by their disability.

In response, some may wonder how we can claim that such individuals have the potential to engage in moral reasoning (by having a conception of their own good and a sense of justice) given they will never be able to do so. I would respond to this by pointing out that if we do not hold that such individuals have, in at least some sense of the word, the potential to engage in this sort of moral reasoning (or indeed any sort of intellectual reasoning) then we cannot make sense of the claim that they are disabled, but that a (non-disabled) dog or chicken, for example, is not. Neither the severely intellectually disabled person nor a dog can engage in abstract intellectual reasoning (which includes moral reasoning), but only one of the two is disabled as a result of this. Why? Because one of them *ought* to be able to do this unlike the other, and one way to articulate this '*ought*' is to say that they intrinsically have the potential to engage in abstract moral reasoning, but that this is being blocked by their disability.

As a result, all human beings are moral persons by merit of being the sorts of things which, when mature and not subject to disability or illness, are capable of having 'a conception of their own good and a sense of justice'. Newborn babies and those with severe intellectual disabilities while not being capable of this in the here and now, are still the sorts of things which, when mature and not subject disability or illness, are capable of doing this, and therefore they too count as moral persons.

We can conclude, therefore, that all human beings will be moral persons at all stages of their life. In this way, there is nothing to stop us combining a political principle of equal consideration of interest with Rawls's account (moral personality) of who counts as a member of our political and moral community.

We can summarize our findings so far in the following principle: 'the political principle of equal consideration of interests for moral persons: when formulating legislation and enacting policy we should give equal weight to the interests of all moral persons who are affected by that legislation, policy, decision-making, and so on'.

Equal consideration for future generations?

We now have a principle with which to tackle our question and we now know to whom it applies. However, it is now worth pausing to note that our question is on *heritable* genomic engineering. It is not just us, those currently existing, who will be affected by this genomic engineering, rather it is us and *future generations* who will be affected. It is worth acknowledging here that some have denied that we can have obligations to future generations. Some have refused to accept, for example, that future generations can be the subject of rights because they do not exist at present and one needs to exist in order to be the subject of anything, including rights.[7] Others have argued that we cannot wrong future people because their very existence is dependent upon the present decisions we make. To give an example, if we destroy the environment now, then a different group of people will come into existence than if we had cared for the environment. It is then unclear how the group of individuals who inherit a degraded environment, and who must suffer hardship as a result, can object given they would not otherwise have existed if we had not destroyed the environment. (If anything, one could argue, they should be grateful since, otherwise, they would never have been born?!) This is the gist of the so-called 'non-identity problem' which was developed by the Derek Parfit (Parfit, 1976).[8] At the same time, numerous responses to these and other similar arguments have been put forward and various different strategies for grounding our obligations to future generation have also been offered.[9]

There is insufficient space to discuss this issue further. As a result, going forward I shall work under the assumption that *we can have obligations to future generations.* I shall make this assumption for two reasons. First, it seems intuitively plausible to claim that we can have direct obligations to future generations, and such a view is widely accepted among contemporary scholars in numerous different contexts (for example, when considering our duty to preserve the environment for future generations). Second, even if we cannot have *direct* obligations to future generations, the political principle of equal consideration of interests is sufficiently flexible so as to provide us with an *indirect* way to ground obligations to future generations via the interests of currently existing people to ensure a good future for their descendants and humanity more generally. Most parents want their children, grandchildren, great-grandchildren, and other descendants, to have a good future. Additionally, most of those who do not have children of their own usually want what is best for their nephews and nieces, godchildren, friends' children, and other human beings in general. As a result, most of us have an interest in the wellbeing of future generations and this is likely sufficient for our purposes. I will illustrate how this might work in greater detail later on.

Genomic engineering and our interests

So far, I have done a number of different things. First, I defended the political principle of equal consideration of interests according to which, when formulating legislation and enacting policy, we should give equal weight to the interests of all members of our political community. Second, I delineated the limits of our political community with reference to Rawls's account of moral personality. Third, I then clarified that the political principle of equal consideration of interests requires us to consider the interests of future generations. This can be done either directly or, if we are suspicious of the idea that we can have direct obligations to future generations, indirectly through the already existing interests of current members of our community in promoting a good future for their descendants and humanity more generally. Thus, we are now in a position to apply our findings to genomic engineering.

I believe there are two distinct ways of understanding the question of this chapter and it is important we do not conflate the two. The first question is: Would heritable genome editing undermine the political principle of equal consideration of interests? With the second question being: What would the political principle of equal consideration of interests say about heritable genome editing? I shall discuss each in turn.

With regard to the first question, it is not clear that heritable genome editing would undermine the political principle of equal consideration of interests. Indeed, there is nothing about heritable genomic editing, *per se*, that would prevent a government from carefully considering the interests of all of its citizens and community members equally.

The only time heritable genome editing would necessarily undermine the political principle of equal consideration of interests would be if it was done in such a way so as to prevent someone being a moral person (that is, having a conception of their own good, and of having a sense of justice). If heritable genome editing resulted in someone not being a moral person, then they would be outside the moral and political community and could be treated accordingly. Whether or not this is an issue in and of itself (after all, they are not moral persons and thus we do not need to consider their interests) is a good question, but it certainly strikes me as dystopian. Would anybody honestly want to edit their genome in such a way so as to render their descendants to be actual sub-persons? Surely not, but it would strike me as unwise to even allow someone to have this choice. This nicely bring us on to the second question.

What would the political principle of equal consideration of interests say about heritable genome editing? This remains unclear because we do not yet know what the limits of heritable genome editing will be. However, I believe it is fair to say it imposes some significant limits on what sorts of heritable genome editing are permitted, and it potentially imposes significant obligations upon the state and society to protect citizens and community members from the negative impacts of heritable genome editing.

Here are a few examples: I have an interest in not having to live alongside a race of supermen whose powers vastly exceed my own. And I have an interest in my children, grandchildren, and so on, not having to live alongside a race of supermen whose powers vastly exceed their own. My grandchildren, who do not yet exist, have an interest in not living alongside a race of supermen whose powers vastly exceed their own. This raises questions about our obligations to future generations as discussed previously. However, as explained, even if we cannot have obligations to future generations then we can simply appeal to my currently existing interests as per my previous point.

If such supermen do come into existence, I have an interest in being protected from them, as do my children, grandchildren, and all my descendants. If heritable genome editing becomes endemic, widespread, and unavoidable, I have an interest in not having the quality of my grandchildren's lives lowered as a result of having (or not having) their genome edited. Additionally, my possible grandchildren, who do not yet exist, have an interest in not having the quality of their lives lowered as a result of having (or not having) their genome edited, and so on.

Returning to the fictional world of *Gattaca*, while genetic discrimination is supposed to be illegal, in reality it is widespread. Clearly the 'invalids' have an interest in having this legislation enforced. Additionally, one of the (sub-)themes of the film is the pressure the 'valids' are placed under. All parents want their children to be 'special': the most athletic, the most intelligent, the most beautiful, and so on. In the fictional world of *Gattaca*, they all try to achieve this through genomic engineering. As a result, no-one is 'special', while, at the same time, all of them feel like they need to be. As a consequence, many of them suffer under the weight of parental expectations resulting in feelings of failure and despair. We can conclude that the fictional world of *Gattaca* is not working for anyone, be they 'valid' or 'invalid'! If the free market and cultural norms are not sufficient to prevent this from happening, then perhaps some form of legislation is needed to prevent these dystopian outcomes.

Yet there are some uses of genomic engineering which, for some, would be permissible, or perhaps even desirable, according to the political principle of equal consideration of interests. For example, the BRCA1 and BRCA2 genes both dramatically increase the risk of breast cancer in women (an 82% lifetime risk) and other genes have been identified with similar effects.[10] This has led many women with these genes to have preventative double mastectomies (De Felice et al, 2015). There is also evidence that there is an increased risk of prostate cancer for male carriers of the BRCA1 and BRCA2 genes (Gorodetska, Kozeretska, and Dubrovska, 2019). It would, therefore, not be unreasonable for a woman with these genes to not only seek to treat them in herself, should that become possible, but also to want to avoid passing these genes on to her children.

Cautions, caveats, and next steps

We can now examine what sort of guidance the political principle of equal consideration of interests provides when applied to genomic engineering. We can see that certain sorts of interventions would be ruled out by this principle, but that other sorts of interventions would be allowed. The details will need to be worked out, but as a rule of thumb I suspect that, as we work through the various possibilities, we will see that interventions to prevent diseases and/or to correct faulty genes will likely be permissible, but that interventions designed to 'improve' otherwise healthy people will likely be impermissible.

However, terms such as health, disease, disability, advantage, and so on, can be tricky to define and, even when they can be defined, they will also depend to some extent on the societal context and other variables.[11] For example, few would consider being shortsighted as a serious disability since glasses are widely available and relatively inexpensive. Yet in an ancient society, without access to these technologies, perhaps it could be considered a disability, albeit perhaps a minor and relatively common one. Similarly, many diabetics may not consider themselves disabled since the condition can be managed with insulin injections with relative ease. Yet, a little over 100 years ago (insulin was first discovered in 1921) diabetes was a serious and potentially life-threatening condition. More contentiously, one could argue that conditions which were not disabilities in the past have become disabilities in the modern world. The modern world is very complex and fast moving. Could we argue that a certain level of intelligence is required to successfully navigate the modern world and that this level might be higher than that required in the past? This is a difficult question, but it does not strike me as obviously false. Thus, perhaps we can argue that in the same way that certain things which are not disabilities in the present were disabilities in the past, certain things which are disabilities in the present were not disabilities in the past.

In the context of genomic engineering this raises a certain difficulty. While we might all agree that women wanting to avoid passing on their BRCA1 or BRCA2 genes to their children is reasonable, what about more ambiguous conditions, such as short-sightedness or mild asthma? Perhaps one could respond by saying we should only modify genes which lead to 'serious' health issues. However, the problem with this response is that 'serious' is subjective and context dependent. If we eliminate only the most serious genetic disorders, then the second most serious genetic disorders will become the most serious. If we then treat those genetic disorders then the third most serious genetic disorders will become the most serious, and so on. Eventually it may well be the case that short-sightedness is considered a serious disability! Is this necessarily an issue? Answering such a question is difficult.

As a result, I think we need to acknowledge that, by itself, the political principle of equal consideration of interest is insufficient to help us fully navigate genomic engineering. We will also need an account of health and disease, and potentially other interrelated concepts. Additionally, as previously stated, we may well want to supplement the political principle of equal consideration of interests with other political principles, such as with an account of autonomy, and human rights.

But there is another issue which requires further consideration. Throughout this chapter when appealing to the political principle of equal consideration of interests, I have assumed that 'interests' are easy to define and that the concept is uncontentious. However, this is far from being the case. We might believe that someone's interests are identical to their desires. If we accept this, however, then the political principle of equal consideration of interests, at least in isolation, loses some of its plausibility. After all, some people want to inflict suffering on others. Is this really an *interest* worth taking seriously in this context? Presumably a great many of us would believe that it is not, but why not? Perhaps we might respond by distinguishing between legitimate and illegitimate interests. We might then argue that a desire to inflict suffering on others is illegitimate in some way. This strikes me as a sensible approach, but clearly more work is needed. What makes one desire or interest 'legitimate' and another 'illegitimate'?

I will not attempt to answer this question here due to limits on space, but I suspect a fruitful avenue of enquiry would be to examine an Aristotelian account of human flourishing (which will likely also give us an account of health and potentially disability). Perhaps we can use such an account to distinguish between legitimate interests (those which accord with human flourishing) and illegitimate interests (those which do not).

It is worth noting that I have no quarrel with the suggestion that there may also be other fruitful lines of enquiry. Here, I am merely offering a suggestion. If the reader is sceptical of Aristotelianism or believes that another line of enquiry is likely to be more promising, then they are free to disagree with me. With regard to this chapter, nothing of weight rests on this suggestion.

Finally, I would like to add a note of caution to what I have written so far. If the history of eugenics has taught us anything, particularly as practiced in the first half of the 20th century, it is that it all too often results in the persecution of those deemed inferior or inadequate. Thus, we must avoid making the same mistakes. More generally, we must be on our guard against the risk of judging the value of people's lives or choices against a perceived ideal. Some individuals may have a genetic disorder and yet they may not wish to 'cure' it. (Perhaps we should not then label it a *dis*order!) They might, for example, see the resulting disability as part of their identity, or they may perceive some advantage to having their disability which outweighs its cost, or they may simply not see it as a disability at all. There are two errors we could make concerning these people. The first, and more obvious mistake, would be to pressure, or even force, them into undergoing genomic engineering to 'treat' their disability. There is insufficient space here to fully justify the claim that it would be immoral to do this. However, at first glance, it seems to me that doing this risks violating their autonomy and authenticity, and that it may initiate a slippery slope towards more totalitarian and dystopian practices.

The second, more subtle, mistake would be to become blind to the additional needs of such people. Persons with disabilities are more than capable of living rich and fulfilling lives and of contributing to society in a meaningful way, but they may need some degree of accommodation in order to do this. For example, while it may not always be caused by a genetic disorder, people in wheelchairs need ramp access to buildings. In the scale of things, this is not an unreasonable ask of the rest of society! As a result, we need to be sensitive to these sorts of needs and we must be willing to accommodate them. There is a risk that, as genomic engineering becomes more common, we will become blind to the needs of people with genetic disorders, and, where we do see them, less willing to accommodate them. Rather than helping persons with a genetic disorder achieve some goal, one could imagine someone simply wondering why their parents, or the persons themselves, did not simply 'treat' the disorder. This may well be an interesting question (and perhaps one a medical professional or good friend of the person with the genetic disorder could ask), but we must not let the fact that such 'disorders' could be 'treated' make us insensitive to their needs (if those are even the right words to use in this context).

Conclusion

In this chapter I have done a number of different things. First, I tried to identify a useful concept of equality that would be both applicable and action-guiding with regard to genomic engineering. I argued that a modified version of Singer's principle of equal consideration of interests, the political principle of equal consideration of interests, fulfilled these criteria. I then clarified to whom this principle is applicable (that is, Rawls's moral persons), before considering how this principle could provide us with either direct or indirect duties to future generations. I then applied the political principle of equal consideration of interests to the issue of genomic engineering. I demonstrated that it provided plausible guidance by allowing certain therapeutic interventions while ruling out more drastic interventions which were at risk, for example, of promoting inequality, promoting discrimination, and undermining the dignity of the human person. Finally, I then added a note of caution on some of the societal dangers that might emerge out of genomic engineering.

In contrast to my previous note of caution, I want to end this chapter on a note of optimism. The human genome has undergone, and is undergoing, a process of constant change. Some of those changes were caused by external pressures, such as the environment (light and dark skin), disease (genetic resistance to certain diseases), and the availability of certain foods (the ability to digest milk). At the same time, some of these changes were caused by internal pressures, most notably sexual selection. Additionally, one could argue that the availability of certain foods is at least partly under our control.[12] On top of this, human beings have always shaped and modified their bodies more directly, for example, by wearing clothes, getting tattooed, and styling our hair.

It is also worth observing that most of these changes to our genome (and phenotype) are either trivial or indifferent in the scale of things (for example, the emergence of blond and ginger hair)[13] or beneficial (for example, resistance to certain diseases). In addition, most of these changes have come about gradually and without a massive disruption to society.

With these insights in mind, we need not see genomic engineering as a completely new and unprecedented stage in human development. We have always tinkered with our genome! The change now is that we can do this in a much more direct, targeted, and rapid manner than we could previously. It must be acknowledged that there are significant risks that come with this ability some of which have been discussed in this chapter. However, provided we are wise, think through these issues carefully and cautiously, and consider issues of equality and justice, among other important principles, there is no reason heritable genomic engineering cannot be a force for good in the world. That strikes me as an encouraging thought!

Notes

1 Our genome is comprised of our DNA. Our phenome, on the other hand, is comprised of our phenotypic characteristics which include all of our physical, measurable characteristics, excluding our DNA. Our phenotypic characteristics are the product of our genes (our DNA), our environment, and the interaction between the two. Examples of phenotypic characteristics would include traits such as the height of a person.

2 As an anonymous reviewer has pointed out, this pattern is not entirely universal. Certain forms of research, particularly in fields like psychiatry and psychology, are now much more tightly regulated than they were in the past. Additionally, some medical interventions, or at least certain forms of said interventions, have been entirely banned in some jurisdictions, for example, so-called conversion therapy is now banned in many countries with many more moving in that direction. As a result, it is not perhaps an entirely foregone conclusion that genomic engineering will be permitted in the future. Only time will tell at this stage.

3 Something is numerically identical only with itself (that is, it is the self-same thing). Two things are qualitatively identical when they have identical features or properties. Two siblings have numerically identical mothers (since they share a single mother), but if each of them is then given an identical bike for their birthday, the two bikes are qualitatively identical.

4 Singer asserts this principle as part of a broader preference utilitarian ethical framework according to which we ought to maximize the preferences of all sentient creatures. Here I am not defending preference utilitarianism. See Sloane (1999) for a criticism of this approach to morality.

5 It should be noted that this general idea is hardly contentious and that it (vastly) predates Singer. It can be identified in the thought of many writers throughout history (certainly all the way back to ancient Greece), and, in our own time, it is a commonly accepted principle in contemporary tax law.

6 It is notable that Singer does not offer a direct quote (or even a specific page number) for Rawls when discussing moral personality.

7 See De George (1981) for an example of this view.

8 See Weinberg (2008) and Woodward (1986) for more detail discussions of this issue.

9 See Meyer (2021) for an overview of this debate, as well as Vicente Bellver's chapter 'Heritable Human Genome Editing and Intergenerational Equality' in this edited volume for a discussion of this issue.

10 See Shiovitz and Korde (2015) for an overview of these issues.

11 See Ereshefsky (2009) for a discussion of health and disease and Kingma (2007) for a discussion of health.

12 For example, see Gibson (2007) for a discussion of the emergence of lactose (in)tolerance and the domestication of cattle.

13 I acknowledge that hair colour is associated with skin colour, and that racism is a serious evil that has long afflicted our species. My point is simply that, *in and of itself*, having blond or brown hair, or light or dark skin, is a trivial matter. The fact that some individuals then practice discrimination and oppression on the basis of these features does not change this basic fact. Indeed, one could argue that it affirms it! One of the reasons racism is so irrational is because it turns what should be a trivial matter into a serious one. This same point can be made about any other form of discrimination on the basis of surface level appearances.

References

De Felice, F., Marchetti, C., Musella, A. et al (2015) 'Bilateral risk-reduction mastectomy in BRCA1 and BRCA2 mutation carriers: A meta-analysis' *Annals of Surgical Oncology*, 22: 2876–2880.

De George, R. (1981) 'The environment, rights, and future generations', in E. Partridge (ed), *Responsibilities to Future Generations: Environmental Ethics*, New York: Prometheus Books.

Ereshefsky, M. (2009) 'Defining "Health" and "Disease"' *Studies in the History and Philosophy of Biology and Biomedical Sciences*, 40: 221–227.

Gattaca (1997) Directed by Andrew Niccol, United States: Columbia Pictures.

Gibson, G. (2007) 'Human evolution: Thrifty genes and the dairy queen', *Current Biology*, 17(8): R295–R296.

Gorodetska, I., Kozeretska, I., and Dubrovska, A. (2019) 'BRCA Genes: The role in genome stability, cancer stemness and therapy resistance', *Journal of Cancer*, 10(9): 2109–2127.

Kingma E. (2007) 'What is it to be healthy?' *Analysis*, 67(294): 128–133.

Meyer, L. (2021) 'Intergenerational justice', in Edward N. Zalta (ed), *The Stanford Encyclopedia of Philosophy* (Summer 2021 Edition), Available from: https://plato.stanford.edu/archives/sum2021/entries/justice-intergenerational/.

Parfit, D. (1976) 'Rights, interests and possible people' in S. Gorovitz et al (eds), *Moral Problems in Medicine*, Englewood Cliffs: Prentice Hall, pp 369–375.

Rawls, J. (1999) *A Theory of Justice*, rev. edn, Cambridge, MA: Harvard University Press.

Singer, P. (2011) *Practical Ethics*, 3rd edn, Cambridge: Cambridge University Press.

Shiovitz, S. and Korde, L.A. (2015) 'Genetics of breast cancer: A topic in evolution', *Annals of Oncology: Official Journal of the European Society for Medical Oncology*, 26(7): 1291–1299.

Sloane, A. (1999) 'Singer, preference utilitarianism and infanticide', *Studies in Christian Ethics*, 12(2): 47–73.

Weinberg, R. (2008) 'Identifying and dissolving the non-identity problem', *Philosophical Studies*, 137(1): 3–18.

Woodward, J. (1986) 'The non-identity problem', *Ethics*, 96(4): 804–831.

Zuolo F. (2017) 'Equality, its basis and moral status: Challenging the principle of equal consideration of interests', *International Journal of Philosophical Studies*, 25(2): 170–188.

PART III

Heritable Genome Editing and the Future

10

Equality of Access to Heritable Genomic Editing

Trevor Stammers

Introduction

Equality of access to treatment for almost any disease or illness is a highly elusive goal. Take malaria for example; the malaria parasite and the mosquito which acts as its vector were both identified in the 19th century. However, effective means of treating the disease have been known since the 17th century, when Jesuit priests first introduced the bark of the American cinchona tree into Europe, though it was not until the 19th century that its active ingredient, quinine, was isolated. Despite the means of prevention and treatment being well-known and established, 90% of the half a million deaths a year from malaria still occur in Africa (Dagan, 2020). If malaria primarily affected Europeans and North Americans,[1] it is likely that it would have been eliminated long ago. We have known of a treatment for malaria since the 1600s, but we still have not succeeded in treating all of those who are currently suffering from it. As a result, we need to be realistic about aspirations of equality of access to heritable genome editing (HGE), should this procedure ever become permitted and effective for the treatment of disease. How much more cautious then should we be about the eventual creation of 'designer babies' often mentioned in media hype about genetic engineering.

Some bioethicists are already more cautious about the possibilities of HGE in the light of what is happening with existing gene therapies, some of which have been available for quite some time. Even somatic genome editing realistically has its limitations in terms of equality of access. Jantina de Vries from South Africa and Françoise Baylis from Canada gently understate that situation of unequal access in relation to sickle cell disease – another condition that is not exactly a priority in the West (de Vries and Baylis, 2021):

> Somatic human genome editing research offers previously unimaginable pathways to alleviating suffering for millions of people living with conditions such as sickle cell disease. Unless specific attention is paid to ensuring the development of affordable innovations that can be implemented across the globe, these benefits may never reach the global poor.

In the UK, use of somatic gene editing treatments for sickle cell disease were only authorized by the National Institute for Heath and Care Excellence as recently as 2025 (Imperial College Healthcare, 2025). The cost for each patient is reportedly £1.65 million (Harwood-Banes, 2025).

Note again that this refers to *somatic* and not heritable genome editing. It is clear that HGE is unlikely to be any less expensive than somatic gene editing and there is an understandable concern that even somatic gene therapy may stall altogether because of its high costs (*Nature* Editorial, 2023). Hemgenix, the first somatic gene therapy licensed by the US Food and Drug Administration for the treatment of haemophilia B, was priced at 3.5 million US dollars per treatment in 2022, making it the most expensive drug in the world at the time (Naddaf, 2022). The ethical imperative of affordable gene editing treatments is a topic of current bioethical discussion (Rueda, Beriain, and Motoliu 2024) but no one yet knows how much such procedures would cost should they be permitted in the future (Cookson, 2023).

The alleged 'imperative' of heritable genome editing

Despite concerns about cost and equality of access, some bioethicists argue that HGE remains not just desirable but a moral imperative. For example, Australian bioethicists Christopher Gyngell, Hilary Bowman-Smart, and Julian Savulescu, making such a case for HGE, conclude (Gyngell et al, 2019, p 522):

> [Natural selection] resulted in vast natural inequality. The most extreme examples are single gene disorders, where some people become destined to a short life with much pain due to random quirks in their DNA. Others are born with high risks of chronic disease like heart disease and cancer. We ought to use powerful technologies like HGE to correct these inequalities and promote human flourishing. Such actions are moral imperatives.

But notice how very far and fast their ground shifts here. They move from somatic gene editing to HGE, from single gene editing to polygenic editing, from existing disorders to *risks* of disorders and from moral choices to moral imperatives – all in the space of one paragraph! Such a rapid escalation invites some probing.

If HGE is taken to be an imperative, then everyone who can do so, presumably, *must* access it. This is a bold claim. But would everyone with even a single gene disorder, let alone a polygenic one, *want* to use it? This seems highly unlikely. For instance, Sir Tom Shakespeare, the UK social scientist, bioethicist, and disability activist, who has achondroplasia – a monogenic disorder, is adamant that he would not have wished his parents to have altered his genes. He writes (Shakespeare, 2015):

> CRISPR–Cas9 is a gene editing tool of great potential, although not necessarily from a disability-rights perspective.[2] People with disabilities are, in my view, unlikely to be queuing up for genetic modification: their priority is to combat discrimination and prejudice.
>
> To 'fix' a genetic variation that causes a rare disease may seem an obvious act of beneficence. But such intervention assumes that there is robust consensus about the boundaries between normal variation and disability. Contrary to the prevailing assumption, most people with disabilities report a quality of life that is equivalent to that of nondisabled people.[3]

When it comes to polygenic disorders, once again, those who live with these conditions do not all appear to want to prevent them. Consider American geneticist George Church, one of the foremost champions for HGE, who has narcolepsy – a polygenic rather than a monogenic condition, which is unlikely to be curable by HGE. He candidly admitted: 'I probably wouldn't fix my narcolepsy even if I could … It illustrates that we need to be very cautious about eliminating certain genes' (Corbyn and Church, 2015). Perhaps then 'ableism', the view that disability is an abnormality instead of a normal feature of human diversity, can lead to flawed 'solutions' and end up disempowering those most affected.[4]

Whatever benefits may result for any further individuals to receive HGE, applying the brake on the likely *adverse* consequences for society as a whole will be far too late, once the procedure is widely permitted.

Equality and inequality of access to HGE, does not begin with the advent of a perfected technique but much further back down the track. Equality of access policies of any new genetic advance do not suddenly appear out of the blue in liberal democracies. Such policies and legislation are always preceded by position papers and opinion pieces from prominent think-tanks which are usually followed by opinion polls, public consultations, citizens' juries and assemblies. Many of the major reports on HGE from various organizations are discussed in other chapters of this volume but I will briefly touch on three of them here, specifically in regard to equality of access.

Aspects of equality of access in key reports on HGE

The 2000 Report from the American Association for the Advancement of Science entitled *Human Inheritable Genetic Modifications: Assessing Scientific, Ethical, Religious, and Policy Issues* mentions matters concerning equal access several times. In the section on social justice, it states:

> There is concern that the technology will enable us to enhance offspring in socially desirable and competitive ways, thereby further privileging the wealthy and powerful by securing the position of their offspring against competition. (American Association for the Advancement of Science, 2000, p 31)

These authors also considered that 'unequal access to HGM (Human inheritable germline modification) technologies would mean that those persons who can already provide the best "environments" for their children would also be able to purchase the best "natures"' (American Association for the Advancement of Science, 2000, p 38).

The UK's Nuffield Council on Bioethics report in 2018 entitled *Genome Editing and Human Reproduction: social and ethical issues* only refers to equality (in any context) a dozen times in its 205 pages and one of those is only to point out that it believes that 'the Netherlands is a technologically advanced liberal democracy with among the lowest measured levels of social inequality' (Nuffield Council on Bioethics, 2018, Section 3.52). The report concludes, as one of its main principles, that HGE 'should be permitted only in circumstances in which it cannot reasonably be expected to produce or exacerbate social division or the unmitigated marginalisation or disadvantage of groups within society' (Nuffield Council on Bioethics, Section 3.68). However, the manner in which such restrictions would be ensured is not elaborated.

The 2017 US National Academy of Science, Engineering and Medicine report *Human Genome Editing: Science Ethics and Governance* (National Academy of Science, 2017) considered, in the context of determining what should and should not be allowed, that the concept of 'fairness' has at least two interpretations. The first is that HGE could be used to create a level playing field for those whose traits put their children at a disadvantage. The second interpretation is that HGE 'should not be permitted to confer a social advantage beyond that which an individual possesses by fate or through personal effort'.

Beyond disease and disability: future inequalities of access to HGE

As mentioned earlier, some people with heritable disability would not have wanted their parents to have utilized HGE to ensure another child was generated instead of them. But should it eventually be permitted and become generally available, at least some parents with genetic disability or disease, will understandably want to access the procedure for their children. But so too will many people with no disability or disease but are affluent enough to afford such modifications in order to ensure their children are selected for enhanced abilities rather than eliminating disability.

I mentioned earlier how Gyngell et al's paper on the alleged 'moral imperative' of HGE for improving human health merges the use of HGE for monogenic disease to polygenic disease and then onto *risk* of disease. But the possible future applications these authors envisage does not stop there. In an earlier paper, Gyngell, Savulescu, and others have already given support for permitting HGE for procedures such as the prevention of ageing, improving educational aptitudes, and for human enhancements, should these prove possible in the future. They indicated: 'Ageing kills 30 million every year and disables many more. Genetic engineering has already produced Methuselah mice that live twice as long as unmodified mice (Bartke et al, 2001). Ultimately, they suggest, gene editing could be used to delay or turn off ageing in humans' (Savulescu, 2015, p 478).

But not all bioethicists see *themselves* as the arbiters of the future of HGE. In her book *Altered Inheritance*, Baylis declares she does not 'want to live in a world where a select privileged few are able to inscribe their privilege in their DNA and thereby exacerbate unfair class divisions and other social injustices' (Baylis, 2019a, p 8). She believes that 'unequal access to genome-editing technologies will both accentuate the vagaries of the natural lottery and introduce an unjust genetic divide that mirrors the current unjust economic and social divide between rich and poor individuals' (Baylis, 2019a, p 67).

Instead, Baylis argues for a broad social consensus as the best way to 'harness the power of genome editing for benefit of humankind' (Baylis, 2019a, p 8). While this is both an understandable and commendable view in a liberal democracy, it should be remembered that not all states in the world subscribe to democratic values and some at least, will do whatever they can, including the use of force, to exert their power over their own and other peoples.

In addition, it seems highly probable that the distinction between health improvements and human enhancements (which is already unclear)[5] will become even more difficult to defend, should HGE eventually deliver its promise, since these two goals are already so intertwined. Such a likelihood certainly gives credence to the American social medicine scholar, Eric Juengst's suggestion that:

> [I]nstead of trying to come to consensus on incommensurable community worldviews and personal values, perhaps it would be better to prepare for a world in which gene-edited enhancements and

> occasional inadvertent germline changes are a reality and to discuss the human rights protections that the variegated inhabitants of such a world will need. (Juengst, 2017, p 21)

Contrary to Juengst's aspiration, surely the problems that societies are already having with their existing 'variegations' ought to sound caution about his view and keep the focus on agreeing limits on HGE before its use if permitted? Juengst, however, is convinced we should 'reorient our conversation from policing science to governing society and shift our focus from avoiding risks to protecting opportunities' (Juengst, 2017, p 22).

But protecting opportunities is arguably a far more nebulous, and therefore difficult, task than policing science. Furthermore, it envisages a universal acceptance of inequality by arranging society so that parents are allowed to enhance their children, though at the same time limiting them from gaining societal advantages. The existing structural inequalities, even within 'advanced' countries like the US[6] and most of Europe,[7] make this outcome very unlikely. Introducing legislation that restricts the genetic modifications parents can make to their children would likely be far easier to implement and enforce from the outset.

Intergenerational changes and equality of access

Looking even further into possible futures, consideration must be given to how equality of access may change over many generations and what problems and challenges may arise in the passing of the centuries.

Racial tensions

Should it become possible to change skin colour through HGE, Baylis raises the question of whether African couples, in the US and much of Europe, would feel it desirable, or even obligatory, to modify their children's skin colour to improve their life chances and life expectancy (Baylis, 2019b). The same could be said of first nation peoples in Australia and New Zealand. What Baylis does not explicitly raise is the spectre of HGE being used by a totalitarian state to *compel* people to become white over the generations or even for this trend to occur through individual 'choice' in the light of ongoing targeted societal oppression. Whether a world in which racial bigotry and hatred prevailed in such a way, would be a better world to live in, is highly doubtful.

Educational issues

To take a totally different example from ethnicity, equal access for all to enhance their children's learning ability may become increasingly meaningless in terms of achieving equality of outcomes as generations progress. Philosopher, Robert Sparrow, points out that any enhancements achieved by genetic manipulation of embryos will offer their parents only a 'one shot' chance at achieving some particular level of enhancement which will be based on the level of knowledge of HGE at that specific time (Sparrow, 2015, p 235).

It is important to emphasize in relation to polygenic traits such as general cognitive ability and speech development, that there is a great deal of scepticism about the predictive value of even having polygenic indices calculated, let alone editing the genes that may be involved (Kozlov, 2022). A polygenic score summarizes the combined effects of many genetic variants on a trait and *imperfectly* predicts an individual's likelihood of expressing a particular trait.

Testing polygenic scores for embryonic selection for traits, such as cognitive ability and even household income, have been on offer commercially for years to individuals having preimplantation genetic diagnosis. However, what is not made clear to prospective buyers is that polygenic indices are calculated on large samples of people from *different* families. Potential embryos for gene editing will usually all come from the same parents and this reduces the expected gain compared with differences predicated on polygenic scores. Furthermore, virtually all these calculations for polygenic scores are done on white Western populations, meaning that if the parent-clients are, for example, from Asian or African populations, the predictive value may be drastically reduced. In addition, the environmental influences on a particular couple contemplating HGE may be completely different from another couple with a similar score, so predications of completing university education, for example, may be totally misleading.[8]

But supposing, in the future, that a high level of accuracy of outcome of HGE does eventually become possible. If parents choose HGE to improve the educational prospects of their child at any particular point, they may then risk spending the rest of their lives in regret that they did not wait a few additional years and had their child that little bit later. There will be a real danger of what Sparrow dubs 'option regret' for both parents and also for their edited children (Sparrow, 2015, p 242). The American bioethicist, Erik Parens, contrasts this to an emphasis on 'gratitude' or 'acceptance', the concept that it is a necessary condition of a good human life that we acknowledge and cherish the fact that there are some things in life that are beyond our powers to change (Parens, 2005).

If, however, parents reject such 'acceptance' as urged by Parens and insist on edits, the longer such parents delay having their edited children, the better off the desired end-result is likely to be, since better modifications resulting in better traits will then be available. Yet, if the delay is too prolonged, the ability to gestate those children without medical and/or social assistance may no longer be possible as the parents will become too old to conceive without the use of assisted reproduction techniques.

As well as their parents' own regret of the timing of gene selection they imposed on their children, the children themselves may also express a range of emotions very far from the gratitude such parents might have anticipated at the time. Just as there is bullying in schools of those who do not have the latest mobile phones, there may be similar discrimination by their peers of those who do not have the most advanced genomes resulting from the more recent developments of HGE. The only difference in these scenarios is that it would, in theory at least, be possible for children to buy (or steal) the latest phone, but they are stuck with their 'outdated' genomes for the rest of their lives. In such a situation, shame about themselves as well as anger and bitterness towards their parents, are more likely responses than gratitude. Similarly, jealousy and discord among their peers as well as between the genetically 'superior' and their 'inferiors' are highly likely to occur. As is already the case with poorer and richer families, the disparity between the 'gene poor' and the 'gene rich' will both mirror and intensify already existing inequalities.

Sparrow suggests that it might be reasonable to insist that any HGE procedure which cannot be updated should not be available, rather than have to confront the prospect of children spending most of their lives with obsolete edits. But if such updateable edits do become available, it may indeed be rational at the time to adopt them because of the positional advantages they provide. However, it may also be equally rational to prefer to live in a world in which the choice of doing so is not available to anyone, in order to circumvent some of the predictable adverse consequences (Sparrow, 2015, pp 246–247). Is our present world of children from genetically selected embryos any better than before embryo selection became available? There seems little indication that societies which offer HGE will become kinder, more generous, contented, or fulfilled. Though of course this could also happen.

Philosopher Michael Sandel, for example, suggests that any positive genetic trait should be available to all and of benefit to all. He writes:

> If our genetic endowments are gifts, rather than achievements for which we can claim credit, it is a mistake and a conceit to assume that we are entitled to the full measure of the bounty they reap in a market economy. We therefore have an obligation to share this bounty with those who, through no fault of their own, lack comparable gifts. Here, then, is the connection between solidarity and giftedness: A lively sense of the contingency of our gifts – an awareness that none of us is wholly responsible for his or her success – saves a meritocratic society from sliding into the smug assumption that success is the crown

> of virtue, that the rich are rich because they are more deserving than the poor. (Sandel, 2007, p 91)

Perhaps, as Sparrow suggests and Sandel urges, those born after us who will have superior intellects, will take care to avoid treating us as their inferiors and will support us financially when we become unable to support ourselves. Alternatively (or additionally), there may be a 'trickle down' of the wealth generated by those with later forms of HGE, to those whose edits have become obsolete (Sparrow, 2015, p 247). However, Sparrow also does wryly add that the wealthy citizens of California already have approximately twice the life expectancy of poor residents of sub-Saharan Africa, yet there is insufficient concern in California about this fact to motivate any meaningful action to address it (Sparrow, 2015, p 247).

Collective action problems and individual benefits

Another area of concern about equality of access to HGE is that of '"collective action problems", where one option is optimific from the perspective of an individual but results in collective harms if everyone pursues it' (Gyngell et al, 2019, p 518). These authors present a hypothetical situation where HGE can be used to select for or against extroversion as a personality trait. They point out that extroversion is highly heritable and linked with increased levels of subjective wellbeing. This, combined with evidence that mothers value extroversion in their children more than if their children were for example, more conscientious or even more highly intelligent, suggests that were parents to have access to HGE to increase their chances of having an extroverted child, they may well decide to take it up. However, as the authors then inconveniently point out:

> As extroversion is associated with higher levels of subjective well-being, such a change would be consistent with a health benefit. But if HGE were to increase the rate of extroverts in society, it is possibly that this would make society worse off as a whole. Introverts contribute important forms of cognitive diversity which can benefit for example group problem solving. Having introverts in society can thus benefit many areas of life including achievement in the sciences and the arts. (Gyngell et al, 2019, p 518)

They also postulate that if parents are able to use HGE to alter characteristics such as height, this could lead to yet more collective action problems. Though each individual child may gain social advantage, if too many people choose these changes for their children, it could make things worse for society as a whole. For example, Gyngell and others indicate that: 'If the average height of the population increased, this could increase the amount of resources that we use, and thus damage the environment' (Gyngell et al, 2019, p 518). Sparrow believes that such collective action problems must mean '[T]hat freedom of access to HGE for all is not absolute and may properly be restricted if the resulting phenotypic changes would lead to harm to others—including if having the freedom to enhance oneself generates a destructive collective action problem' (Sparrow, 2015, p 251).

However, Gyngell et al prioritize another issue which they consider to be a much greater priority with which this chapter began:

> It is clear that the use of modern medicine is worsening the position of members of future generations, by allowing random mutations to occur to our genome. Fortunately, there is a straightforward compensatory action—developing HGE. This is not something that is merely permissible—but a moral imperative. (Gyngell et al, 2019, p 522)

All the advantages of modern medicine have, in their view, made things worse for future generations, meaning that HGE for all then becomes a moral imperative to put everything right again. Sparrow, however, is much more sobering, and maybe realistic, in his assessment when he concludes that 'Genetic enhancements pose a particular danger of an enhancement rat race because it seems highly unlikely that they will be capable of being updated and because they seem especially likely to be closely associated with social status' (Sparrow, 2015, p 253).

Continuing progress in enhancing health by HGE may, in time, turn out as something to be deeply regretted rather than celebrated, if the enhanced were the only ones who could ever excel. Unless of course it was excelling in kindness and self-sacrificial service to others. But if we lived in a world where that was the case many (though not all) of the reasons for supporting the idea of a moral imperative for HGE would disappear.

Notes

1. Which may of course happen if predictions about its future epidemiology in the light of climate change prove accurate. See for example Beck-Johnson (2013). Even so, in Northern and Western Europe the risk of malaria may not rise significantly if the current loss of wetlands due to increasing urbanization continues to eliminate the mosquitos' breeding habitat. See Piperaki et al, 2016.
2. Shakespeare reference here is Mathews et al, 2015.
3. Shakespeare reference here is Albrecht and Devlieger, 1999.
4. See, for example, Wolbring, 2014.
5. See, for example, Erler, 2017 and Juengst et al, 2018.
6. See, for example, Amadeo, 2022.
7. See, for example, Boke, 2021.
8. For a comprehensive and detailed review of these issues, see Harden, 2022.

References

Albrecht, G.L. and Devlieger, P.J. (1999) 'The disability paradox: High quality of life against all odds', *Social Science & Medicine*, 48: 977–988.

Amadeo, K. (2022) 'What is structural inequality? How structural inequality stifles the American dream', *The Balance*, 26 March 2022, Available from: www.thebalancemoney.com/structural-inequality-facts-types-causes-solution-4174727# [Accessed 16 October 2024].

American Academy for Advancement of Science (2000) 'Human inheritable genetic modifications: Assessing scientific, ethical, religious, and policy issues', Available from: www.aaas.org/sites/default/files/germline.pdf [Accessed 15 October 2024].

Bartke, A. et al (2001) 'Extending the lifespan of long-lived mice', *Nature*, 22(6862): 412–414.

Baylis, F. (2019a) *Altered Inheritance: CRISPR and the Ethics of Human Genome Editing*, Cambridge, MA: Harvard University Press.

Baylis, F. (2019b) 'Human genome editing: our future belongs to all of us', *Issues in Science and Technology*, 35: 42–44.

Beck-Johnson, L.M. et al (2013) 'The effect of temperature on anopheles mosquito population dynamics and the potential for malaria transmission', *PLoS ONE*, 8(11), Available from: https://journals.plos.org/plosone/article?id=10.1371/journal.pone.0079276 [Accessed 1 September 2023].

Boke, S.S. (2021) 'Socio-economic inequalities in Europe: time to restore social trust by strengthening social rights', Strasbourg: Council of Europe, Available from: https://assembly.coe.int/LifeRay/SOC/Pdf/TextesProvisoires/2021/20210909-SocialInequalities-EN.pdf [Accessed 16 October 2024].

Cookson, C. (2023) 'Cost of gene treatments for once incurable diseases too high, scientists warn', *Financial Times*, 8 March 2023.

Corbyn, Z. (2015) 'George Church: 'Genome sequencing is like the internet back in the late 1980s'', *The Observer*, 18 February 2015, Available from: www.theguardian.com/science/2018/feb/18/professor-george-church-nebula-genomics-interview [Accessed 15 October 2024].

Dagan, M. (2020) 'History of malaria and its treatment', in G.L. Patrick (ed), *Antimalarial Agents. Design and Mechanism of Action* (e-book), pp 1–48, Available from: www.sciencedirect.com/science/article/pii/B9780081012109000019

de Vries, J. and Baylis, F. (2021) 'Equity and access need to be at the forefront of innovation in human genome editing', *The Conversation*, 12 July 2021.

Editorial (2023) 'The gene-therapy revolution risks stalling if we don't talk about drug pricing', *Nature*, 616: 629–630.

Erler, A. (2017) 'The limits of the treatment-enhancement distinction as a guide to public policy', *Bioethics*, 31(8): 608–615.

Gyngell, C., Bowman-Smart H., and Savulescu J. (2019) 'Moral reasons to edit the human genome: Picking up from the Nuffield report', *Journal of Medical Ethics*, 45: 514–523.

Harden, K.P. (2022) *The Genetic Lottery: Why DNA Matters for Social Equality*, Princeton: Princeton University Press.

Harwood-Baynes, M. (2025) '"Groundbreaking" sickle cell treatment to be offered to NHS patients for first time', Available from: https://news.sky.com/story/groundbreaking-sickle-cell-treatment-to-be-offered-to-nhs-patients-for-first-time-13299986 [Accessed 14 July 2025].

Imperial College Healthcare (2025) 'World-first gene-editing treatment for blood disorders including sickle cell disease approved for use in the NHS', Available from: www.imperial.nhs.uk/about-us/news/gene-editing-for-sickle-cell-approved-for-use-on-the-nhs [Accessed 10 July 2025].

Juengst, E.T. (2017) 'Crowdsourcing the moral limits of human gene editing?', *Hastings Center Report*, 47(3): 15–23.

Juengst, E.T., Henderson, G.E., Walker, R.L., Conley, J.M., MacKay, D., Meagher, K.M., Saylor, K., Waltz, M., Kuczynski, K.J., and Cadigan, R.J. (2018) 'Is enhancement the price of prevention in human gene editing?', *CRISPR Journal*, 1(6): 351–354.

Koslov, M. (2022) 'The controversial embryo tests that promise a better baby', *Nature*, 6 October 2022, Available from: www.nature.com/articles/d41586-022-02961-9

Mathews, D.J.H. et al (2015) 'CRISPR: a path through the thicket', *Nature*, 527: 159–161.

Naddaf, M. (2022) 'Researchers welcome $3.5-million haemophilia gene therapy — but questions remain', *Nature*, 612: 388–389.

National Academy of Science, Engineering and Medicine (2017) 'Human genome editing: Science ethics and governance', Available from: https://nap.nationalacademies.org/catalog/24623/human-genome-editing-science-ethics-and-governance [Accessed 15 October 2024].

Nuffield Council on Bioethics (2018) 'Genome Editing and Human Reproduction: social and ethical issues', Available from: www.nuffieldbioethics.org/publications/genome-editing-and-human-reproduction [Accessed 15 October 2024].

Parens, E. (2005) 'Authenticity and ambivalence: Toward understanding the enhancement debate', *Hastings Center Report*, 35: 34–41.

Piperaki, E.T. and Daikos, G.L. (2016) 'Malaria in Europe: Emerging threat or minor nuisance?', *Clinical Microbiology and Infection*, 22(6): 487–493.

Rueda, J., Beriain, I.M., and Motoliu, L (2024) 'Affordable pricing of CRISPR treatments is a pressing ethical imperative', *The CRISPR Journal*, 7(5): 220–226, https://doi.org/10.1089/crispr.2024.0042

Sandel, M.J. (2007) *The Case Against Perfection: Ethics in the Age of Genetic Engineering*, Cambridge, MA: Harvard University Press.

Savulescu, J., Pugh. J., Douglas, T., and Gyngell, C. (2015) 'The moral imperative to continue gene editing research on human embryos', *Protein Cell*, 6(7): 476–479.

Shakespeare, T. (2015) 'Gene editing: Heed disability views', *Nature*, 527: 446.

Sparrow, R. (2015) 'Enhancement and obsolescence: Avoiding an "enhanced rat race"', *Kennedy Institute of Ethics.*

Wolbring, G. (2014) 'Ability privilege: A needed addition to privilege studies', *Journal for Critical Animal Studies*, 12(2): 118–141.

11

Heritable Human Genome Editing and Intergenerational Equality

Vicente Bellver

Introduction

Whenever heritable human genome editing (thereafter, HHGE) is discussed, future generations are also discussed. This is because very few individuals exist who have resulted from this procedure[1] but it always affects future generations, that is, human beings who do not yet exist. As the 2020 Geneva Statement on Heritable Human Genome Editing indicates, 'Heritable human genome editing would not treat, cure, or prevent disease in any existing person. Instead, it would modify the genes of future children and generations through the intentional creation of embryos with altered genomes' (Andorno, 2020).

In this chapter, I will try to answer the question whether the interests of future generations have any relevance when deciding if heritable human gene editing should be allowed.[2] This will be based on a twofold presumption: (1) that in the not-too-distant future the thresholds of safety and efficacy, which are considered sufficient to proceed with the creation of new human beings, will be reached, and (2) that by the time the technique may be safe, it will be impossible to stop its uses, not only for the suggested purpose of preventing very serious monogenic diseases in future children, but also for other uses.

This may happen, firstly, for moral reasons. In other words, if the technology to 'correct' these 'genetic defects' is developed, many will believe that we may have a duty to use it, and not just to correct but to enhance capacities (Veit, 2018). Secondly, for pragmatic reasons, insofar that if the technology becomes safe and effective, it may become available to many. This would also imply that the argument for rejecting HHGE on the grounds of avoiding increased inequality would be neutralized.

My claim in this chapter is that HHGE should not be carried out precisely to protect the interests of future generations. In this regards, it may be useful to distinguish between three different groups of future persons: (1) those produced as a consequence of a heritable genetic modification; (2) those born to people created through a heritable genetic modification, who may or may not be aware of the genetic modification that affects their own genetic heritage; and (3) those in future generations in the long term, who would be significantly affected by the deliberate modification of humanity's genetic heritage. With some exceptions, in the following discussion, I will not address the specific issues of intergenerational justice that arise between the current generation and each of the future generations mentioned. Instead, the considerations being addressed below are of a general nature and can be taken into consideration in relation to all three groups of persons.

Three main arguments may exist for rejecting HHGE. First, a prudential argument that calls for caution in introducing changes whose future impact is unpredictable. Second, a strong argument, but one that will not meet with broad consensus, that such interventions, insofar as they damage the genetic integrity of future generations, lead inexorably to an undesirable alteration of human identity. Indeed, behind this attempt to enhance the human genome may lie a very questionable conception of the human being and the world as it appears to us, namely Gnosticism. And, third, an argument based on consequences, which pays attention to the collateral effects of eventual uncontrolled competition: The fear that money and genes work together to produce a society marked by eugenics and genetically increased social inequalities, generating what could be called 'genobility' (Green, 2008). Science fiction stories may already warn us about some of these risks. More than a century ago, the English writer H.G. Wells (1866–1946) presented, in his novel *The Time Machine* (1895), the human race separated into two species. And his compatriot Aldous Huxley (1894–1963) described in a *Brave New World* (1932) a society divided in five classes according to their different genetic abilities.

Many argue that this opposition to HHGE is wrong from a moral perspective and 'naive' from a sociological perspective. They suggest that it is morally wrong to prohibit a technique that does not cause harm to third parties and can ensure that future human beings are born without serious genetic problems (Savulescu, 2007). Some go much further and argue that there is a moral duty to improve the genetic characteristics of our offspring (Veit, 2018). As HHGE may be able to provide great benefit to the individuals of the future, society may express a strong demand for it when it is proved safe.

However, it should be noted that the achievement of a significant benefit to future individuals without causing harm either to the direct beneficiaries or to others can never be fully, or even sufficiently, guaranteed. It would therefore be gravely imprudent to trigger an action whose effects are unforeseeable and could be extremely damaging to future generations. Consequently, society's initial enthusiasm in the face of such promising expectations could lead to a strong rejection and, above all, to enormous confusion as to how to proceed in a future in which the genetic modification of the human species has failed and it is impossible to return to the past.

A starting point: international declarations on human rights

So far, UNESCO has adopted two important declarations to guarantee human rights in relation to the human genome: the Universal Declaration on the Human Genome and Human Rights (1997, endorsed by General Assembly of United Nations in 1998) and the International Declaration on Human Genetic Data (2003). The first does not contain an express reference to the need to preserve the integrity of the human genome for future generations. But it does make an indirect reference, stating that 'the International Bioethics Committee of UNESCO... should make recommendations, in accordance with UNESCO's statutory procedures, addressed to the General Conference and give advice concerning the follow-up of this Declaration, in particular regarding the identification of practices that could be contrary to human dignity, such as germ-line interventions' (Article 24). The Declaration on Human Genetic Data refers to future generations by warning of the impact that genetic data may have on them (Article 4).

The importance of giving due regard to the impact of genetic technologies on the genetic constitution of future generations was specifically emphasized by Article 16 of the Universal Declaration on Human Rights and Bioethics, which was approved by the General Conference of UNESCO in 2005. This 'due regard' is not only justified by the risks related to safety or abuse of the technology. Rather, the authors of the Declaration were concerned that humanity would be seriously harmed if its genetic constitution was altered (Morisaki, 2009, pp 143–149).

The notion that we may have responsibilities towards future generations has already been clearly established in international documents regarding the protection of the environment. For instance, the 1992 Rio Declaration on Environment and Development states: 'The right to development must be fulfilled so as to equitably meet developmental and environmental needs of present and future generations' (Principle 3). The same terms were used in the Vienna Declaration and Programme of Action of the World Conference on Human Rights (1993). In section 11 it indicates: 'The right to development should be fulfilled so as to meet equitably the developmental and environmental needs of present and future generations'.

Although it is difficult to interpret such ambiguous legal texts, the Rio Declaration itself and also the United Nations Declaration on the Right to Development (1986) provide some clues for the interpretation of the above-mentioned principle. Principle 1 of the Rio Declaration states that 'human beings are at the center of concerns for sustainable development. They are entitled to a healthy and productive life in harmony with nature'. The Declaration on the Right to Development proclaims the right to development as an inalienable human right by virtue of which every human person and all peoples are entitled to participate in and contribute to (Article 1). It states that all human beings 'have a responsibility for development, individually and collectively, taking into account … their duties to the community, which alone can ensure the free and complete fulfilment of the human being' (Article 2).

According to these articles, intergenerational justice must guarantee the development of present and future generations. This is understood as a right and a duty; the holders of the right to development are both individuals and peoples; the community is essential to achieve development; and the goal of development is a life in harmony with nature. Therefore, when evaluating our interventions in nature (such as the deliberate alteration of humanity's genetic heritage as a consequence of HHGE), we should recognize that we not only have an individual right but also a collective right. In other words, that we have a duty to contribute to this collective right by fulfilling our responsibilities towards the community, and that this right must seek a harmonious relationship with the nature of present and future generations. Therefore, in accordance with the spirit of the aforementioned standards approved by the United Nations, any individualistic approach focused solely on the present, and ignoring the importance of nature, must be discarded. The double dependence of human beings on nature and on the community is not seen as an obstacle to overcome but rather as the path to achieving full development (Bellver, 2023).

The Maastricht Principles approved in 2023 are fully aligned with this approach. They do not include any explicit reference to HHGE but the Preamble states:

> Humanity, the Earth on which we live, the natural systems of which we are but one part, and our political, social, cultural and economic systems, are in the midst of profound, rapid, and perilous change at humanity's own hands (V).

Adding: 'The rights of future generations must be interpreted and applied in light of humanity's dependence on and responsibility to Earth's natural systems, now and throughout our species' future' (XIII).

The relevance of nature

The idea of humanity's relationship with nature supposes living in a way that respects and aligns with the natural world, which is something that we are born into and that helps us grow and thrive. Nature, in this sense, is a guiding normative principle that encompasses both what is outside human beings as well as our own biological constitution (MacIntyre, 1999). Of course, this vision of nature does not lead to enshrining biochemical and ecological processes as laws to which we must inexorably submit our lives and actions. To do so would be to commit an unacceptable naturalistic fallacy. The concept of nature to which I appeal is based on the following statements: (1) Nature is part of reality; (2) Nature can be known by human beings; (3) Nature has a meaning or purpose; (4) Nature offers us a moral horizon common to all human beings (Spaemann, 1987); (5) This moral horizon is specified in general guidance principles when discerning the correct action that we must take in each situation; (6) This discernment is not the conclusion of logical-deductive reasoning but the result of a process of deliberation, weighing and argumentation; (7) The concept of nature as a moral reference, has as its fundamental objective the realization of justice, which consists of safeguarding the universal dignity of the human being; (8) Currently, this safeguard is basically contained in the International Bill of Human Rights, which includes the Universal Declaration of Human Rights of 1948 and the International Covenant on Economic, Social and Cultural Rights, and the International Covenant on Civil and Political Rights, both adopted in 1966; (9) According to nature, well-constituted societies are those that guarantee what is fair for all human beings, not those that impose a certain vision of what is good (Ignatieff, 2003); (10) To determine the fundamental content of human dignity that must be safeguarded in all cases, it is essential to assume a conception of the human being, which we discover by knowing nature and, specifically, the nature of the human being (Ballesteros, 2001).

This conception of nature as a normative horizon is at the foundation of the main declarations of rights (Morsink, 2009). For example, the Declaration of Independence of the United States begins with a powerful reference to nature and its moral laws:

> When in the course of human events, it becomes necessary for one people to dissolve the political bands which have connected them with another, and to assume among the powers of the earth, the separate and equal station to which the Laws of Nature and of Nature's God entitle them.

Also, the Virginia Declaration of Rights (1776) states: 'That all men are by nature equally free and independent and have certain inherent rights'

(1776). The French declarations of the rights of man of 1789 and 1793 do not mention nature, but do mention natural rights. It is true that neither the Universal Declaration nor the International Covenants include any reference to nature. But they all discuss the 'inherent dignity of all members of the human family' (Morsink, 1999).

It is, of course, no secret that the concepts of nature and dignity, understood as the foundation of human rights, have been the subject of intense controversy (Waldron, 2015). For this reason, some authors consider them useless and propose their replacement with the concepts of autonomy or respect (Macklin, 2003). But before the unanimous appeals to nature and human dignity found in the main declarations of human rights are dismissed as ambiguous, rhetorical, or ideological, it seems logical to investigate the reasons for this unanimity. It could turn out that, by trying to replace or eliminate these references, the entire architecture of human rights and intergenerational justice, as it is currently embodied, would collapse.

The shadow of Gnosticism

One interpretation of Article 16 of the Universal Declaration on Human Rights and Bioethics is coherent with the corpus of texts on human rights but is based on a vision of our relationship between humans and nature that is opposed to the dominant view, namely misguided anthropocentrism (Bellver, 1994; Ballesteros, 1995). Such anthropocentrism denies that human beings can find guidance for their lives in nature because they only see nature as a collection of raw materials to exploit through their technological ingenuity (Kopnina et al, 2018). It also dismisses the human biological condition as radically flawed. This ideology has very deep roots. Over the centuries it has appeared at different times and with great appeal, as witnessed through the evolution of Gnosticism (Jonas, 2001).

According to this school of thought, the world is not a wonderful divine creation but the perverse work of a demiurge, in which humanity has been imprisoned. Only a few, chosen to access authentic knowledge (gnosis), are able to discover the deception and escape from it by overcoming the perverse material world. In Western culture we can trace Gnosticism back to Hellenic culture and early Christianity. Some authors consider that Modernity finds one of its foundations in Gnosticism (Voegelin, 1952).

However, today's neognosticism is not inspired by a spurious interpretation of Christianity but by what German philosopher Günther Anders calls the 'Promethean shame and the obsolescence of human beings' (Anders, 1952). According to this author, the human being feels ashamed of himself because he feels that the artifacts that are the result of his ingenuity are superior to him. Consequently, he strives to produce new human beings who are as perfect as the things he produces.

Throughout history, Gnosticism has manifested itself in two contrasting ways. On the one hand, it is a contempt of the material world while seeking, instead, luminous mystical-spiritual experiences. On the other, it is a desire to manipulate matter without limit so that it satisfies all the desires of the individual. Contemporary neognosticism opts for this second path, and there it meets scientism (Rey, 2003). The confluence of both leads to the complete negation of the meaning of everything that human beings receive as 'given': The living and mortal body; the nature that provides for all our needs and enjoyment but requires our care; procreation as a way of giving life to new human beings, etc. (Ballesteros, 2021).

One of the most prominent representatives of this philosophy is the British philosopher John Harris. His position is based on what was already defended by techno-enthusiasts in the 1970s (Gray, 2004): From the moment science offers human beings the chance to guide their own biological evolution, it becomes their obligation to do so. Leaving aside the problems of safety and abuses, Harris holds that a germline intervention does not pose any special ethical problems because, like education, it is nothing more than an instrument to enhance our children:

> Now suppose, as is much more likely, we could use genetic engineering, regenerative medicine or drugs, or reproductive technology or nanotechnology to produce healthier, fitter and more intelligent individuals. What should our reaction be? Would it be unethical to do so? Would it be ethical not to do so?
>
> Our question is this: if the goal of enhanced intelligence, increased powers and capacities, and better health is something that we may strive to produce through education, including of course the more general health education of the community, why should we not produce these same goals, if we can do so safely, through enhancement technologies or procedures? (Harris, 2010, p 2)

The rhetorical questions posed by Harris prove extremely persuasive. But closer analysis of the reasoning behind these proposals may reveal inconsistencies. For example, enhancement procedures, and in particular heritable genome editing, cannot be used to achieve enhanced intelligence, health, or another capacity in a particular person, for the simple reason that they cannot be achieved by these means. It only creates a human being who is different from all those who have previously existed. Who can know what intelligence or health means for a person who is different from us? But even if we maintain that this difference is of little consequence, who can guarantee health, enhanced intelligence, or other capabilities by means of manipulating genes?

Heritable genome editing only generates certain kinds of children. In this case, the progenitors are the ones who control the genetic heritage of possible future children. Even if done with the best of intentions, they inevitably break with the essential symmetry between the generations as regards how human beings have been conceived until now (Habermas, 2003). The progenitor's role is replaced by the genetic 'controller', who must assume responsibility for the genetic heritage of the possible future child. The genetic characteristics of the child will be unalterable. But can the end subjects (the children) take legal action if they are not satisfied with the genetic heritage imposed on them? Under what circumstances? And against whom can they make their claim?

More fundamentally, education and germline gene enhancement are completely different and even opposite activities. The purpose of education is to assist individuals achieve their full potential, and this goal can only be attained by allowing individuals to exercise their freedom in the educational activities in which they are involved. In other words, education cannot exist without the voluntary participation of the individual. In contrast to education, the genetic enhancement of future individuals is imposed on them; they cannot give any consent because, by definition, they do not even yet exist. Hence, it is deeply misleading to appeal to educational purposes to justify germline human enhancement.

When germline interventions are used to improve the intelligence, health, or abilities of future individuals, we are not really educating them. In fact, we are doing the opposite of what education does, which enables students to reach their goals by empowering them to freely exercise their abilities. Heritable genome editing, on the other hand, would force upon a child certain traits that the progenitor believes would enhance him or her, without any input or choice from the future child.

It is true that education includes the development of instrumental capacities (memory, calculation, logical reasoning, etc.), some of which may be enhanced more effectively by germline intervention. But the aim of education is not to develop superlative instrumental capabilities in the individual for their own sake. Quite the contrary, these are only considered to the extent that they are necessary for a flourishing life. While germline intervention imposes enhancements that the subject has not decided for her/himself and cannot be modified, education provides these enhancements as and when the individual needs them.

Of course, we make all manner of choices for our (present and future) children based on what we believe may be best for them. But it could be argued that making decisions about their genomes is different, more personal, and restrictive than making decisions about their environment.

Finally, behind the use of hereditary genome editing is the idea that we know what is best for our descendants. By altering the genes of those to come, we dare to project our limited vision of the world onto others. Just as in the past we colonized many territories convinced that it was best for the peoples we invaded, today some persons want to colonize the future of our descendants because they are convinced that they know what is best for future generations (Innerarity, 2009; Krznaric, 2021).

Faced with this vision that spurns nature, it is possible to maintain that the most important normative instruments on human rights seek to preserve the human species as well as nature because they implicitly acknowledge that these two goods are necessary for human existence and flourishing, for 'a life in harmony with nature'.

At present it is commonly accepted that nature is not a storehouse of raw materials to be exploited as we see fit. This vision, dominant during a good part of modernity, has been discredited by the deep ecological crisis that we are now experiencing caused by the uncontrolled power of certain technological innovations. We were convinced that we were going to be capable of doing something much better than nature but what we achieved was putting it at risk. We were unaware of having a power whose effects on nature, other individuals, and the future, was unpredictable. This unbridgeable distance between our action and its effects prevents us from adequately understanding the immense power of intervention over the future that we have acquired and allows us to carry out truly dangerous actions as if we were doing an ordinary job (Anders, 1988).

It is important to understand that the way we would treat humanity if we implemented HHGE would be similar to the way we treated nature in the past, which resulted in disastrous consequences. We may believe that we are able to create better human beings, without knowing the answers to the questions: What are the standards to determine what is 'better'? Better for whom? Better for what? Perhaps one might answer that it is for the benefit of human beings. The problem is that once we have discarded human nature as a normative standard, we can no longer determine what is 'best' for humans. It is inconsistent that those who defend the possibility and even the duty of overcoming as well as improving the human condition (transhumanism), or even a substantially different human condition (posthumanism), deny the existence of a guiding normativity of human action (Marcos and Pérez Marcos, 2019). If human beings are deficient, what reason is there to believe that they can achieve something better than themselves?

The imperative to protect nature and humankind

Surprisingly, there are far more international regulations demanding the protection of nature than the preservation of humanity. The UNESCO International Bioethics Committee rightly pointed out this inconsistency when it claimed: 'Intergenerational obligations have mostly been considered in environmental policy, with limited focus on the effects of technological advances in the life sciences and their impact on future generations' (UNESCO International Bioethics Committee, 2021, n. 172).

One of the most important and symbolic international rules on this matter is the 1997 UNESCO Declaration on the Responsibilities of the Present Generations Towards Future Generations, because it discusses the need to simultaneously preserve nature and humankind. According to Article 3, 'the present generations should strive to ensure the maintenance and perpetuation of humankind with due respect for the dignity of the human person. Consequently, the nature and form of human life must not be undermined in any way whatsoever'. Similarly, Article 6 provides that 'the human genome, in full respect of the dignity of the human person and human rights, must be protected and biodiversity safeguarded. Scientific and technological progress should not in any way impair or compromise the preservation of the human and other species'.

These articles (and the entire Declaration) are based on the conviction, deeply held among many individuals of all cultures that what we have received is more important than what we can project. First, because the existence that allows us to develop a life in freedom has not been created by ourselves but has been given to us. Second, because what is given is not ours but belongs to all present and future humanity. And third, because what is given is the main source of inspiration and the basis for our flourishing. As the famous quote goes (attributed to many individuals) 'We do not inherit the Earth from our ancestors—we borrow it from our children'. When we read 'earth' we can also include humankind.

The above-mentioned report on the principle of the protection of future generations, published by UNESCO's IBC in 2021, reaffirms that 'Genetic technologies, particularly germline intervention, are of great concern due to their implications for future generations', and emphasizes some important points for our discussion:

- That most national and international institutions have strongly discouraged or prohibited germline interventions on ethical grounds.
- That States should not become involved in germline interventions because, according to Article 24 of the Universal Declaration on the Human Genome and Human Rights, it could be contrary to human dignity.
- That the 1997 European Convention on Human Rights and Biomedicine (also known as Oviedo Convention) also prohibits the introduction of any modification in the genome of descendants (Article 13).

On the issue of equality in relation to future generations, the IBC report offers interesting reflections, such as in Paragraph 80. This reflects the reality that responsibility for future generations is abstract because it concerns human beings we will never know, who will live in a future far from ours, and whose individual and collective choices may be very different from ours. It is therefore difficult to identify the aspirations of these future generations. But precisely because we cannot identify the aspirations of future generations it is essential not to interfere in the genetic constitution of our offspring.

As the German philosopher, Markus Gabriel, has pointed out, the undeniable difference between us and other animals (that humans are minded living beings who seek to understand the world and themselves and who possess ethical insight into moral contexts) defines the human condition and places a special responsibility on us to consider our actions in relation to ourselves as human beings, to other living beings, and to our shared habitat. But it also calls on us to cultivate an ethics of not-knowing: to recognize that, however much we may seek to understand the world, we will never completely master it. Our grasp of reality will always be limited: much is, and will remain, alien to us, lending itself only to speculation. Remembering this will help us to preserve humanity's identity and the habitability of the Earth in the face of the immense power that technology has placed in our hands and the existential risks it can unleash (Gabriel, 2024). An ethics of not-knowing would include Hans Jonas´ imperative of responsibility as its stone corner: 'Act so that the effects of your action are compatible with the permanence of genuine human life' (Jonas, 1984).

What to do?

We have seen that, depending on the conception we assume about the relationships between humans and nature, this will influence our position on the future use of biotechnologies and, in particular, HHGE. In a context in which it may be impossible to reach universal agreement on this point, we can consider the proposals of some of the international bioethics committees that have dealt with this issue and that are fully aware of the plurality of opinions on the matter.

In 2021, the European Group of Ethics (EGE) in Sciences and New Technologies published a document on human heritable genome editing, entitled Opinion on Ethics of Genome Editing, which presents an insightful consideration about the possibility of using heritable genome editing for avoiding heritable disorders. It points out that:

> [T]here appears to be general agreement that germline genome editing, hence introducing heritable changes, is not to be applied at this point. In many fora have its potentially severe risks – for the individuals concerned and for society overall – been discussed. Together with the difficulty to conduct long-term studies and the availability of alternative methods for avoiding heritable disorders, they require us to ask: Are research on embryos and the risk of harm caused by the technology

> ethically acceptable and proportionate for the few cases for which there is no alternative solution? Questions like these require broad and well-informed societal deliberation on the basis of an awareness about how heritable genome editing may result in major changes of a society overall, its composition and its values. (European Group on Ethics in Science and New Technologies, 2021, p 6)

Similarly to the EGE report, the Declaration of the Organising Committee of the Third International Summit on Human Genome Editing (London, 2023) also stresses the importance of public discussions and policy debates because they are important in resolving whether this technology should be used. The statement establishes a series of minimum conditions, although not necessarily sufficient, that should be met in order to allow any case of this type of genetic editing: 'Heritable human genome editing should not be used unless, at a minimum, it meets reasonable standards for safety and efficacy, is legally sanctioned, and has been developed and tested under a system of rigorous oversight that is subject to responsible governance. At this time, these conditions have not been met' (UK Royal Society et al, 2023). It is difficult to foresee how the first and third requirements can be met in the future: How is it possible to guarantee the safety of a technique whose effects are propagated from generation to generation, and how is it possible to ensure that, in the long term, this technology is developed under rigorous supervision.

Years before, in 2015, the American Society for Gene and Cell Therapy and the Japanese Society of Gene Therapy published a joint position statement on human genome editing. It probably contains one of the most prudent positions delivered for a scientific society on this matter. They recognize that gene editing in human embryonic cells or in stem cells destined for use in the creation of a human being may, in the long run, have the potential for correcting genetic defects associated with genetic disease, not only in patients, but theoretically in future generations as well. But they are also quick to acknowledge that this field of research is complex.

The requirement that the results of an experiment be susceptible to analysis and characterization before further applications are undertaken cannot be met with human germ-line modification with current methods, because the results of any such manipulation could not be analysed or understood for decades or generations – a situation incompatible with ethical imperatives and with the scientific method (Friedmann et al, 2015).

As with the EGE report and the Declaration of the Third Summit, they propose to develop effective social and policy mechanisms for carrying out broad and deep discussions of human germ-line alteration in order to effectively understand and balance the individual, familial, societal, and perhaps even species-level rights, needs, interests, and values affected by this rapidly advancing science. But this American–Japanese document went beyond what those statements propose, concluding that 'these ethical and safety concerns are serious enough to support a strong stance against gene editing or genetic modification of human cells to generate viable human zygotes with heritable germline modifications' (Friedmann et al, 2015).

Final remarks

In light of the preceding discussion, it is possible to briefly summarize the following considerations:

1. Although many intergovernmental organizations or international bioethics committees have expressed great concern about the serious risks of HHGE, and indicated that this technology does not offer sufficient guarantees for its use for reproductive purposes at this time, their statements and guidelines are being interpreted in very ambiguous ways.
2. There is a need for an open public debate on the regulation of this technology given that its consequences would affect the future of all humanity. Only after such a public deliberation can legitimate decisions be taken in this area. Unfortunately, the current debate on HHGE is seen as being monopolized by a scientific and financial minority which generally represents private interests.
3. It is highly unlikely that the procreative freedom of future parents will be limited so that HHGE is only used to avoid the birth of children with very serious monogenic diseases. At best, only when a HHGE procedure, itself, is seen as being unsafe, resulting in the possible birth of a future child with a disorder, would it be prohibited.
4. Unfortunately, only if catastrophic situations for the future of humanity are likely (as is currently the case with climate change), and after extensive use of this technology, will there be a possibility for reconsidering what was done (but not before). But it may then be too late. However, this possibility may be unlikely because the negative effects of this technology may be more difficult to perceive, as they will take place gradually.
5. Beyond imagining and assessing the possible effects on the first generation of individuals born as a result of heritable genome editing, it may be useful for society to reflect on three essential aspects of the future of humanity, namely:
    - Do we want technology to be a tool for individuals to help develop their lives or for it to become the ultimate goal that guides our lives?
    - Do we want human time to be subject to the laws of biology and human narratives or to be supplanted by an increasingly accelerated succession of meaningless instants? (Han, 2023).

- Do we accept that human beings are inevitably vulnerable, finite, and interdependent? Or do we aspire to turn them into perfectly self-sufficient and invulnerable individuals?

Depending on how we answer these questions, we may either have a post-human future, with some similarities to Huxley's *Brave New World*, or a human future, with the lights and shadows of humanity as in the past.

6. The international norms dedicated to our obligations towards future generations aim to protect nature and preserve humankind. While there is a broad consensus on the first objective, there is an intense debate on the scope of the second objective.
7. The most prudent thing we can do for future generations when we consider using HHGE is to give ourselves time. Despite the pressures, there is no real urgency to develop this technology and many concerns remain to be considered.

Notes

1 Currently, there is only evidence of the birth of two twin girls (Lulu and Nana) and a third girl (Amy) resulting from HHGE in China (Cyranoski, 2019).

2 The argument that persons who do not yet exist may have 'rights' is controversial and regarded by legal scholars as conceptually flawed. Probably for this reason, no international legal instrument, to date, refers to the 'rights' of future generations, but rather to our duties towards future generations. This means that humankind is an 'object' of our responsibility, not a 'subject' of rights. For instance, the 1997 UNESCO Declaration on the Responsibilities of Present Generations Towards Future Generations does not mention the 'rights' of future generations.

At that time, UNESCO asked a committee of legal experts to consider whether it was appropriate to talk about such 'rights' and the answer from the experts was negative. This is why the Declaration does not refers to such 'rights' in any Article. Despite this opinion, a group of international law experts began to work in 2017 on a project which culminated in the adoption of the Maastricht Principles on the Human Rights of Future Generations at an Expert Seminar hosted by Maastricht Centre for Human Rights on 3 February 2023. These Principles represent an initial attempt to elaborate in detail the implications of regarding future generations as human rights-holders under international law (Suárez Franco and Liebenberg, 2023).

References

Anders, G. (1952) *Die Antiquiertheit des Menschen Bd. I: Über die Seele im Zeitalter der zweiten industriellen Revolution*, Munich: Beck.

Anders, G. (1988) *Wir Eichmannsöhne. Offener Brief an Klaus Eichmann*, München: Beck.

Andorno, R. et al (2020) 'Geneva statement on heritable human genome editing: the need for course correction', *Cell*, 38(4): 351–354, Available from: www.cell.com/trends/biotechnology/fulltext/S0167-7799(19)30317-8 [Accessed 25 October 2024].

Ballesteros, J. (1995) *Ecologismo Personalista*, Madrid: Tecnos.

Ballesteros, J. (2001) *Sobre el Sentido del Derecho*, 3rd edn, Madrid: Tecnos.

Ballesteros, J. (2021) *Domeñar las Finanzas, Cuidar la Naturaleza*, Valencia: Tirant lo Blanch.

Bellver, V. (1994) *Ecología: de las Razones a los Derechos*, Granada: Comares.

Bellver, V. (2023) *Derechos al Final de la Vida*, Madrid: REUS.

Cyranoski, D. (2019) 'The CRISPR-baby scandal: what's next for human gene-editing', *Nature*, 566: 440–442.

European Group on Ethics in Science and New Technologies (2021) 'Opinion on the ethics of genome editing', Available from: https://op.europa.eu/en/publication-detail/-/publication/6d9879f7-8c55-11eb-b85c-01aa75ed71a1 [Accessed 25 October 2024].

Friedmann, T., Jonlin, E.C., King, N.M.P., Torbett, B.E., Wivel, N.A., Kaneda, Y., and Sadelain, M. (2015) 'ASGCT and JSGT joint position statement on human genomic editing', *Molecular Therapy: the Journal of the American Society of Gene Therapy*, 23(8): 1282, Available from: www.cell.com/molecular-therapy-family/molecular-therapy/fulltext/S1525-0016(16)30162-9 [Accessed 25 October 2024].

Gabriel, M. (2024) *The Human Animal: Why We Still Don't Fit into Nature*, Cambridge: Polity.

Gray, J. (2004) *Heresies: Against Progress and Other Illusions*, London: Grant Books.

Green, R.M. (2008) 'Are babies by design in our future?', *Personalized Medicine*, 5(3): 273–277.

Habermas, J. (2003) *The Future of Human Nature*, Cambridge: Polity.

Han, B.Ch. (2023) *Die Krise der Narration*, Berlin: Matthes & Seitz.

Harris, J. (2010) *Enhancing Evolution. The Ethical Case for Making Better People*, Princeton: Princeton University Press.

Ignatieff, M. (2003) *Human Rights as Politics and Idolatry*, Princeton: Princeton University Press.

Innerarity, D. (2009) *El Futuro y sus Enemigos*, Barcelona: Paidós.

Jonas, H. (1984) *The Imperative of Responsibility: In Search of an Ethics for the Technological Age*, Chicago: University of Chicago Press.

Jonas, H. (2001) *The Gnostic Religion: The Message of the Alien God and the Beginnings of Christianity*, Boston: Beacon Press.

Kopnina, H. et al (2018) 'Anthropocentrism: More than just a misunderstood problem', *Journal of Agricultural and Environmental Ethics*, 31: 109–127.

Krznaric, R. (2021) *The Good Ancestor: How to Think Long Term in a Short-Term World*, London: WH Allen.

MacIntyre, A. (1999) *Dependent Rational Animals*, Chicago: Open Court Publishing.

Macklin, R. (2003) 'Dignity is a useless concept', *BMJ (Clinical Research Ed.)*, 327(7429): 1419–1420.

Marcos, A. and Pérez Marcos, M. (2019) 'Caverna 2.0. Las raíces dualistas del transhumanismo', *Scientia et Fides*, 7(2): 23–40.

Morisaki, T. (2009) 'Article 16: Protecting future generations', in H. ten Have and M. Jean (eds), *Universal Declaration on Human Rights and Bioethics: Background, Principles and Application*, Paris: UNESCO.

Morsink, J. (2009) *Inherent Human Rights: Philosophical Roots of the Universal Declaration*, Philadelphia: University of Pennsylvania Press.

Morsink, J. (1999) *The Universal Declaration of Human Rights: Origins, Drafting, and Intent*, Philadelphia: University of Pennsylvania Press.

Rey, O. (2003) *Itinéraire de l'égarement. Du rôle de la science dans l'absurdité contemporaine*, Paris: Le Seuil.

Savulescu, J. (2007) 'In defence of Procreative Beneficence', *Journal of Medical Ethics*, 33: 284–288.

Spaemann, R. (1987) *Das natürlische und das Vernünftige*, München: Piper Verlag.

Suárez Franco, A.M. and Liebenberg, S. (2023) 'The Maastricht principles on the human rights of future generations', in J. Martens (ed), *Spotlight on Global Multilateralism Perspectives on the future of international cooperation in times of multiple crises*, Bonn: Global Policy Forum Europe, pp 59–65, Available from: www.globalpolicy.org/sites/default/files/Spotlight_Multilateralism_web.pdf [Accessed 25 October 2024].

UK Royal Society, UK Academy of Medical Sciences, US National Academies of Sciences and Medicine, and The World Academy of Sciences (2023) 'Statement from the Organising Committee of the Third International Summit on Human Genome Editing', London, Available from: https://royalsociety.org/-/media/events/2023/03/human-genome-editing-summit/statement-from-the-organising-committee-of-the-third-international-summit-on-human-genome-editing.pdf [Accessed 25 October 2024].

UNESCO International Bioethics Committee (2021) 'Report on the principle of protecting future generations', Paris, Available from: https://unesdoc.unesco.org/ark:/48223/pf0000378723 [Accessed 25 October 2024].

Veit, W. (2018) 'Procreative beneficence and genetic enhancement', *Kriterion – Journal of Philosophy*, 32(1): 75–92.

Voegelin, E. (1952) *The New Science of Politics. An Introduction*, Chicago: University of Chicago Press.

Waldron, J. (2015) 'Is dignity the foundation of human rights?', in R. Cruft, S.M. Liao and M. Renzo (eds), *Philosophical Foundations of Human Rights*, Oxford: Oxford University Press.

12

Heritable Genome Editing, the Future of Equality, and the Need for Social Justice and Human Rights Perspectives

Katie Hasson

Introduction

This chapter sets out to explore realities of and aspirations to social equality, both present and future, as they relate to heritable genome editing. I approach this question as a social scientist and civil society advocate, stressing the need to understand heritable genome editing in context – including how social structure, cultural biases, and power hierarchies would shape its development, marketing, and use. Doing so orients the conversation to the ways in which present-day inequalities would shape the use of heritable genome editing in ways that diminish prospects for future equality. I argue for the inclusion of civil society advocates in deliberations on heritable genome editing as a way to ensure that equality stays at the forefront of the conversation.

Social justice advocates have been largely missing from high-level, 'official' discourses on heritable genome editing, such as the major reports from national science academies and the three international 'Summits' on human genome editing. But scholars and advocates committed to social justice and human rights have been conducting their own deliberations, where the conversation looks much different. This chapter describes one such set of deliberations convened by my organization, the Center for Genetics and Society, to draw attention to concerns about heritable genome editing from the perspectives of disability, gender, LGBTQ, racial and reproductive rights and justice, health equity, human rights, and Indigenous sovereignty.

Heritable genome editing raises many risks to equality and the values that flow from it. The procedure would fundamentally change the relationship between generations by allowing parents to attempt to control or even design their children's genetic heritage. It would threaten universal human rights by undermining the idea that all people possess equal social worth from birth (Darnovsky, Lowthorp, and Hasson, 2018; MacKellar, 2021). By allowing those who could afford it to alter their children's genes and traits, heritable genome editing would exacerbate existing social inequalities, which could lead to a new, market-based techno-eugenics. This is particularly true if heritable genome editing were to be used to attempt enhancements.

Genetically enhancing complex traits like intelligence is unlikely to work, but even purported genetic enhancement would harm our commitment to equality by designating some people as biologically more desirable from birth. Some of heritable genome editing's strongest proponents note this potential to undermine equality and human rights (although not all see it as a problem) (Silver, 1998; Buchanan et al, 2000).

Some proponents have argued that these dangers could be mitigated by using lotteries, subsidies, or other means to ensure that everyone has equal access to heritable genome editing (Salkever and Wadhwa, 2017; Macintosh, 2018). This is not only extremely unlikely, but also beside the point. Within a society rife with inequality, heritable genome editing would most likely be used in ways that reinforce the advantages of those at the top of the social hierarchy – even if it were made widely accessible. Understanding how heritable genome editing might affect the future of equality, therefore, requires turning our attention to present-day inequality.

The voices and perspectives of civil society advocates are instrumental in making this shift. Advocates with lived experience and expertise on inequality and oppression can be best situated to see how emerging technologies, if they are developed and deployed in an inequitable context, would intensify social inequality and be particularly harmful to marginalized communities.

While calls for the public to play a role in decisions about heritable genome editing have been made by individual scientists, national science academies, and international bodies, what they envision differs widely, from surveys to public empowerment programmes (Jasanoff and Hurlbut, 2018; Baylis, 2019b; Andorno et al, 2020; Darnovsky and Hasson, 2024). But many in the scientific and bioethics communities, particularly those that have taken the lead in high-profile science policy conversations, seem committed to limiting the role of the public in determining the future of heritable genome editing.

A full accounting of the risks of heritable genome editing to prospects of equality in society requires that we broaden, not narrow, public engagement and include civil society perspectives grounded in social justice and human rights. What does the case against heritable genome editing look like when we centre social context and consequences through the voices and perspectives of scholars and advocates committed to social justice and human rights?

Changing the conversation about heritable genome editing

A first step in changing the conversation about heritable genome editing would be to examine where it has gone wrong. Distortions and misconceptions about heritable genome editing have had the effect of making meaningful public engagement more difficult. When the terms of the debate put concerns about equality out of bounds, it works to exclude civil society voices that are best equipped to discuss heritable genome editing's effects on equality. 'The Geneva Statement on Heritable Human Genome Editing' is a call for reorienting the public discourse, written by 'public interest advocates, social science and humanities scholars, ethicists, policy experts, and life scientists who share a commitment to social justice, human rights, and democratic governance of science and technology'. The authors argue:

> In order to have the societal debate about heritable genome editing that almost everyone claims to support, we need to identify the shortcomings of the discourse thus far and correct a number of misunderstandings and distortions that have taken hold. Success in that endeavor would remove one set of barriers to fair and genuine deliberations. (Andorno et al, 2020)

Discussions about the acceptability of heritable genome editing tend to focus on technical issues and the potential benefits or harms at an individual level. In other words, it is treated like other medical interventions (National Academy of Medicine, National Academy of Sciences, and The Royal Society, 2020). But heritable genome editing is not a medical treatment. Its effects are aimed at conceiving a certain kind of human being, and therefore it does not treat anyone or prevent anyone from getting sick (Rulli, 2019; Mills, 2024). In most existing techniques, CRISPR would be applied at the moment when egg and sperm are combined in the laboratory, meaning no embryo exists before gene editing occurs.

Somatic gene editing, in contrast, changes DNA only in targeted cells in the treated person's body and, in theory, does not affect future generations. Promising gene therapies have recently been approved, and many more are in development. Meanwhile, the very high prices of these first treatments are presenting a real affordability crisis for individuals, insurers, and health systems. But it is essential not to conflate these therapies with heritable applications: somatic gene therapies treat patients; heritable editing aims to alter the traits of future children and generations (Baylis and de Vries, 2021; WHO Expert Advisory Committee, 2021).

Some argue in favour of heritable genome editing on the grounds that it would allow prospective parents to avoid passing on a genetic condition to future children. But safer and less socially fraught alternatives already exist. In addition to adoption, using third-party eggs or sperm can ensure an unaffected child. Preimplantation genetic diagnosis (PGD), which is an embryo screening technique, allows prospective parents, who know they are at risk of passing on a genetic condition, to select unaffected embryos. Embryo selection raises serious ethical concerns, as it involves making choices about which lives will be welcomed into the world. But making heritable genetic changes would greatly amplify those concerns (MacKellar, 2021; Darnovsky and Hasson, 2024).

What heritable genome editing actually offers is the opportunity to have fully genetically related children unaffected by a particular genetic condition for those few people for whom PGD would not work. The strong and compelling desire to have fully genetically related children is not, however, a medical need. Therefore, deliberations about heritable genome editing must consider a more comprehensive range of concerns that goes beyond individual risks and benefits – including ethical, cultural, and social implications.

Understanding this distinction makes space for a broader public conversation. When heritable genome editing is framed in technical terms and promoted as a medical treatment, it is scientists, doctors, and bioethicists who are granted the most authority in the conversation. But if the issue is reframed in terms of parental desires, weighing these against the risks to society allows the conversation to engage a more nuanced and robust consideration of social equality (Baylis, 2019a; Rulli, 2019; Mills, 2024). This, in turn, opens up the range of expertise and experiences that are considered relevant to the discussion.

Another approach to the question of heritable genome editing argues that there are certain acceptable uses that should be allowed. These are variously described as 'therapeutic' or 'medical' applications restricted to 'serious disease' – and contrasted with 'non-medical' or 'enhancement' uses. This approach is problematic for several reasons. First, there is no agreement on what constitutes a 'serious' disorder (Wertz and Knoppers, 2002; Kleiderman, Ravitsky, and Knoppers, 2019). Second, most countries lack policy mechanisms for restricting the use of medical technologies in this way. Further, this approach assumes that societal questions, like equality, apply only in cases of enhancement.

An example of this approach can be found in the 2020 report from the International Commission on the Clinical Use of Human Germline Genome Editing. This committee, convened and appointed by US and UK science academies, was explicitly tasked to develop a 'translational pathway' to clinical use of heritable genome editing (National Academy of Medicine, National Academy of Sciences, and The Royal Society, 2020). In its recommendations for that pathway, the report specified that the first clinical uses should be limited to cases where prospective parents would have no (or poor) choices for having a genetically related child who would be unaffected by a serious monogenic disease (National Academy of Medicine, National Academy of Sciences, and The Royal Society, 2020, p 100).

The report said little about *how* the use of heritable genome editing would be limited to specific cases. Policy mechanisms to do so are lacking in many if not most countries. Physicians in the United States, for example, are permitted to prescribe drugs or devices for 'off-label' uses beyond the specific FDA-approved indications. This could allow eager providers in the under-regulated fertility industry to use heritable genome editing for any purpose, even if the initial approval was quite narrow (Charo, 2019).

Finally, the idea that heritable genome editing could be permitted for some uses but not others downplays the existing policy reality. There are 70+ countries with policies that categorically prohibit heritable genome editing (Baylis et al, 2020). This includes 31 countries that have ratified the Oviedo Convention, a binding international treaty from the Council of Europe that prohibits heritable genome editing on the basis of its threat to universal human rights and human dignity. The Oviedo Convention has been in place since 1997, and the Council of Europe recently reaffirmed the prohibition of heritable genome editing in the face of arguments that it should be lifted (Council of Europe, 1997; Genome Editing Technologies, 2021). This striking degree of agreement among countries has gone under-acknowledged in high-level policy discussions, which approach the topic of global governance as if starting with a blank slate.

Broadening, not narrowing, public deliberation

Broadening the scope of values under consideration and making space for civil society voices would require a 'course correction' in conversations about heritable genome editing (Andorno et al, 2020).

Re-orienting the conversation towards heritable genome editing's societal context and consequences would shed light on how this technology would interact with existing inequalities that could affect prospects for equality in the future. Doing so means considering heritable genome editing not in a vacuum, but as embedded in, and deeply informed by, the cultural, social, historical, and commercial contexts that would shape its development and use. This context includes the commercialization of reproductive technologies within the global fertility industry, eugenic legacies and temptations, and the impacts heritable genome editing could have for marginalized groups.

This contextualized approach makes it clear that a broader range of expertise, beyond science and academic bioethics, is essential to a full consideration of the issues. Filling in the gaps and blind spots of expertise, narrowly defined, requires thinking differently about whose expertise is recognized and valued, as well as seeking expertise across a range of academic disciplines – and beyond academia. This must include civil society advocates whose lived experiences and political work provide clear views of how this powerful new technology would diminish prospects for equality.

There is extensive agreement that some form of public participation on the question of whether to pursue heritable genome editing is necessary (Sarewitz, 2015; Burall, 2018; Jasanoff and Hurlbut, 2018; Baylis, 2019b; Lander et al, 2019; Andorno et al, 2020; Dryzek et al, 2020). Despite this, the role imagined for the public's participation seems to be shrinking. In 2015, the influential statement from the organizers of the first International Summit on Human Gene Editing made a powerful call that placed 'broad societal consensus' on equal terms with proven safety and efficacy as prerequisites for using heritable genome editing (Baltimore et al, 2015).

Since 2015, the aspirations towards broad societal consensus have been siphoned off and watered down. Following the Second International Summit in 2018, the leaders of the US, UK, and China's national science academies published a commentary in *Science* that proposed establishing a new international commission to chart a 'responsible' pathway that would bring heritable genome editing to the clinic. This was despite having joined the global condemnation of disgraced scientist He Jiankui mere weeks prior. In doing so, they shifted the marker on what kind of agreement would be needed to pursue heritable genome editing:

> [National Science] Academies are well-positioned to convene needed international expertise and to help foster *broad scientific consensus* on the responsible pursuit of human genome editing research and clinical applications. (Dzau, McNutt, and Bai, 2018, emphasis added)

The replacement of 'broad societal consensus' with 'broad scientific consensus' went hand in hand with the statement that scientists were best positioned to make determinations about whether heritable genome editing should be pursued (Baylis, 2019b; Baylis and Darnovsky, 2019).

In 2019, the announcement of a new committee convened by the US and UK science academies proceeded in a similar vein:

> An international commission will be convened … to develop a framework for considering technical, scientific, medical, regulatory, and ethical requirements for germline genome editing, should society conclude such applications are acceptable. (International Commission

on the Clinical Use of Human Germline Genome Editing | National Academies, no date)

In this one sentence, the gesture toward 'society' having the last word on whether 'such applications' should be developed and used is already undercut by the call to lay out the pathway. By the time we would get to 'society's' decision (mentioned at the end of that long sentence), there would not be much left to decide. The statement may appear to allow 'society' to place limits on scientific research and technological developments in human genome editing. In reality, it both ignores the limits that have already been placed (by 70+ countries with existing prohibitions and numerous human rights instruments) and undercuts its own allowance that any limits would be accepted.

When the international commission published its 2020 report *Heritable Human Genome Editing* and laid out a fully-fledged technical pathway for bringing heritable genome editing to the clinic, bioethicist Peter Mills of the UK Nuffield Council pointed out:

> The pathway is a powerful organising metaphor for technology. The pathway drives traffic to its destination. If you have not asked for a pathway, or if you don't want a pathway, being presented with fully drawn-up plans for a pathway may seem a little presumptuous. (Mills, 2020)

The 2020 report was conspicuously silent about what resources and mechanisms might be available for empowering societies to determine whether heritable genome editing should be allowed and hold scientists accountable to that decision. The standard of 'broad societal consensus' was further eroded in the call for: 'moving HGE forward thoughtfully and cautiously, only if there is *clear scientific consensus* to continue and only if *a given country decides* to permit its use.' (National Academy of Medicine, National Academy of Sciences, and The Royal Society, 2020, emphasis added)

Not only was responsibility for consensus shifted from society to scientists, but 'society's' role was now delegated to individual countries. Each, it is assumed, will make its own decision – setting the stage for widespread reproductive tourism and 'ethics dumping' (WHO Expert Advisory Committee, 2021).

Civil society voices are essential

Refocusing conversations about heritable genome editing on social context and consequences goes hand in hand with including civil society voices and perspectives to generate a more robust and inclusive public conversation. As noted in the Geneva Statement, 'Public engagement and empowerment are likely to reveal additional concerns that have not yet surfaced, particularly if we commit to including and listening to a broad range of voices and perspectives' (Andorno et al, 2020).

If these voices and perspectives are prioritized, fundamentally different questions and ways of considering the issues emerge. Fostering exchanges among a diverse cross section of civil society would tap into situated, ongoing, concrete conversations about social values and aspirations rather than artificial, isolated, abstract discussions created from scratch.

Modelling public engagement that centres social justice and human rights

At the Center for Genetics and Society, we work with a range of social justice advocacy groups to draw attention to concerns from the perspectives of disability, gender, race, LGBTQ, reproductive rights and justice, health equity, human rights, and Indigenous sovereignty. Together, we consider the likely societal consequences of human genetic and reproductive technologies, assessing them against our aspirations for greater social equality, inclusion, solidarity, and the capacity to democratically shape our collective lives.

Our work with advocates and publicly engaged scholars involves facilitating conversations that take an intersectional approach to examining the societal consequences of genetic and reproductive technologies. Through these conversations, advocates make connections between the values and commitments that drive their work and the often abstract-seeming issues of reproductive and genetic technologies.

While only a few of our partner organizations focus primarily on gene editing or other genetic technologies, they all recognize how a powerful technology like heritable genome editing might threaten hard-won rights and the cultures of disability communities; or how existing sexism, racism, and attempted erasure of Indigenous peoples might shape gene editing's development and commercialization. Advocates from a wide variety of backgrounds can draw on their lived experience and work to build a fair and inclusive society as resources for understanding how heritable genome editing could further curtail prospects for equality. Together, we question the power that is so often ceded to 'science' and 'technology' to shape the future, and to advantage or disadvantage particular communities. In other words, we start the conversation as Francoise Baylis has suggested in *Altered Inheritance* and elsewhere, with the question 'What kind of world do we want to live in?'

These opportunities to engage with other advocates and scholars build capacity among civil society organizations to engage in broader conversations around reproductive and genetic technologies. They also model public engagement that centres social justice and human rights and prioritizes the participation of civil society. These discussions have yielded a range of grave concerns about the impacts of heritable genome editing on society and the conclusion that there is no place for such a technology in the future to which we aspire.

CGS' missing voices initiative

In 2021, CGS launched the Missing Voices Initiative as a model for changing the conversation about heritable genome editing by centering the voices and perspectives of advocates and scholars committed to social justice and human rights. We wanted to highlight how social justice advocates are well positioned to anticipate and identify the difficult and even harmful effects of adding powerful new technologies to inequitable systems. These voices – not actually missing but rather shut out or ignored in high-profile science policy discussions – should be prioritized.

The Missing Voices Initiative began with a push for greater inclusion of civil society representatives at the Third International Summit on Human Genome Editing, which took place in London in March 2023. This initial phase of work culminated in a series of virtual panel discussions and a two-part virtual 'counter-summit' preceding the Summit in London, all featuring advocates and scholars participating in the Missing Voices Initiative.[1] Examples from these discussions on disability rights narratives, intersectional feminism, racial justice, health equity, and indigenous sovereignty are incorporated below to demonstrate just how different the conversation about heritable genome editing is when it begins from questions of equality and justice.

Equity, sovereignty, and racial justice

Mainstream discussions of human genome editing come closest to considering equality when they raise concerns about who will be able to access somatic gene therapies. Even in this case we need to broaden the scope beyond matters of distributive justice. Sky-high prices raise urgent questions about whether those who most need these therapies will be able to afford them. But critical equity concerns arise long before therapies are made available on the market. These include who is at the table setting research priorities, who is set to benefit from commercializing treatments, and whether marginalized communities will be asked to bear the risks of development and testing of potential therapies, with little if any promise of benefit (Baylis and de Vries, 2021; Creary, 2023).

When it comes to heritable human genome editing, focusing on equality of access is unhelpful at best. Heritable genome editing could reinforce dominant ideas about what traits are desirable and encourage the belief that social inequalities are caused by biology rather than social structure. It would all too likely increase inequality and discrimination even if it were somehow made widely available. From the perspective of social justice advocacy, we can take a step back to consider how history, social structure, and power hierarchies would differentially shape outcomes even if equal access could be assured.

The first panel discussion featuring scholars and advocates from the Missing Voices Initiative was entitled: 'Equity, Sovereignty, and Racial Justice: Beyond Access in Debates on Human Genome Editing'. The panel responded to the question, 'How should we talk about equity in the context of human genome editing?'

Panel member Dorothy Roberts argued that social justice and human rights-oriented perspectives provide key insights for 'questioning the deep but often unstated assumptions underlying advocacy for human heritable or germline genetic modification' (Roberts et al, 2022). This questioning requires understanding the legacies of eugenics and how such eugenic ideologies are at play in the present, in part through essentialist understandings of difference. Roberts pointed to the ways that heritable genome editing would reinforce genetic essentialism and reproduce inequality:

> The dominant ethical framework promotes the traits of socially advantaged groups and individual enhancement rather than social change. By assuming a concept of human flourishing that privileges these interests and discounts the importance of social change, arguments in favor of gene editing can reinforce socially unjust hierarchies, even if the technologies are accurate and even if they're widely distributed. (Roberts et al, 2022; see also Roberts, 2019)

Those eager to pursue heritable genome editing 'fixate on what remains to be improved for them, that is, their genomes'. They have 'great faith in the power of changing the biology of individual human beings but little faith in changing the unjust structure of the societies in which human beings live' (Roberts et al, 2022). In fact, focusing on heritable genome editing as a way to improve wellbeing actually draws attention and efforts away from widely recognized social and structural factors that promote equity and demonstrably improve individual and societal wellbeing.

Another critical consideration typically missed in mainstream policy discussions of heritable genome editing is the commercial context in which such a technology would be developed, marketed, and used. Krystal Tsosie, co-founder of the Native BioData Consortium, the first US Indigenous-led biobank, highlighted the disconnect between equity and access in her comments about how Indigenous peoples and other small populations are seen by scientists as valuable to study, a 'treasure trove' of rare (in the broader population) variants. But under a profit-driven model, the value in these variants is based on their potential to benefit the dominant population. It does not extend to developing treatments that would benefit Indigenous populations, precisely because they are too small to constitute a profitable market. She explained:

> When biomarkers from disenfranchised peoples in the Global South have been made openly and publicly available, the entities to benefit first have been drug companies and pharmaceutical companies. The people to benefit last have been the very same Indigenous peoples that provided that data to begin with.
>
> The system of 'benefits most' is one that will continue the further erasure of small populations like Indigenous peoples. So we should be advocating for equity as opposed to democratization, and these two terms are not synonymous. (Roberts et al, 2022; see also Tsosie, 2021)

Underlying these concerns about profit, commercialization, and exploitation is a broader understanding of equality. As the educator and panelist, Milton Reynolds, explained, eugenics and its underlying assumptions about biological superiority were fundamental in giving a scientific sheen to beliefs that some groups were entitled to extract from and exploit others for their own benefit.

Forging new disability rights narratives

Preventing the births of people with disabilities is often used as an argument in favour of heritable genome editing. Underlying these arguments is often a simplistic model of disability that conflates disability and disease, assumes a life defined by suffering, and identifies 'bad genes' as the cause (Garland-Thomson, 2019, 2020). The focus on editing out 'bad' genes often runs contrary to the perspectives of the people and communities who carry those genes.

Disability communities and disability rights advocates have been at the forefront of anticipating and working to prevent the potential harms of heritable genome editing (Generations Ahead, 2010; Burke, 2017; Cokley, 2017). People with disabilities – especially those who do not desire 'cures' – and disability rights advocates should be primary participants in deliberations about heritable genome editing. Instead, they are frequently excluded. When people with lived experience of genetic conditions are included in these discussions, they are most often represented by patient or disease advocacy groups. These organizations tend to share perspectives based on a medical model of disability that aligns closely with scientists' views. They are therefore more often included in discussions around gene editing and other genetic technologies (Generations Ahead, 2010).

In part due to the lack of certain disability rights voices, high-profile science policy discussions often demonstrate little recognition of disability as distinct from illness. For example, in 2017, the US National Academies published a 328-page report on human gene editing that included 82 occurrences of the words 'disability' or 'disabilities'. Of these, 55 were part of the phrase 'disease and disability' or a close variant (National Academies of Sciences, Engineering, and Medicine, 2017). In the most recent 200+ page National Academies report focused on germline editing, the words 'disability' or 'disabilities' appeared only 13 times, and 11 of those – 85% – were part of the phrase 'disease and disability' or a close variant (National Academy of Medicine, National Academy of Sciences, and The Royal Society, 2020; Center for Genetics and Society, 2023). The three-day programme of the 2023 International Summit on Human Genome Editing did not include speakers representing disability rights or disability communities. However, 'patients' and 'patient advocacy groups', which are organized around the experience of seeking medical care and encourage funding for research toward 'cures', were well represented.

In November 2022, CGS organized the virtual panel 'Forging New Disability Rights Narratives on Heritable Genome Editing', which demonstrated how the perspectives of disability studies scholars and disability rights advocates enrich the debate. The discussion placed heritable genome editing in the context of histories of eugenic sterilization in the US and the exclusion of disabled children from education. It highlighted how ableism and implicit bias among medical professionals contribute to people with disabilities having medical treatments denied to them in some cases and imposed on them in others (see also, Tsaplina and Stramondo, 2020; Lagu et al, 2022; Meredith et al, 2023).

Panelist Larkin Taylor-Parker, Legal Director of Autistic Self-Advocacy Network, expressed frustration that, when disabled people are included in discussions about heritable genome editing, they:

> [S]truggle to get [the conversation] past the threshold question of whether our lives are worthwhile, whether it's a good thing that they occur in the world in the first place. Often it is so difficult for us to convince people outside of our communities that there is something that we offer to the world, that our lives have quality and meaning and value, that we can never clear that threshold question and delve into the risks of heritable genome editing and the reasons there might be for the widespread consensus that it's simply too dangerous. (Yee et al, 2022)

This aspect of the conversation can take a toll, and it is therefore important to be thoughtful about what is being asked of people with disabilities. It is also important that these conversations include disabled people with a range of views and give them the power to determine the scope of the conversation, so that they are not conscripted into narrowly defined conversations with pre-determined ends that effectively 'use' them to add legitimacy to proponents' arguments. But the current lack of representation of disabled people in scientific and medical professions as well as in policy circles, particularly among decision-makers, leaves too much room for ableist assumptions to fill in for those 'missing voices'.

Panelists also challenged the explicit or implicit assumption that prospective parents – those with and without disabilities – would always be motivated to avoid having a child with a disability. They pointed out disability communities that value and welcome multiple-generation families, as in Deaf, little people, and autistic communities. These families are celebrated for the richness of shared experiences, culture, and resources that contribute to individual and community flourishing.

If allowed, heritable genome editing could increase stigmatization of people with disabilities, diminish necessary social supports and resources, and even threaten the existence of thriving disability cultures and communities. New narratives that are generated within disability communities, which value disabled people's lives and rich cultures, can counteract the false assumptions, ableism, and eugenic legacies that support many arguments in favour of heritable genome editing.

Feminist intersections

CGS' 2023 webinar, 'Feminist Intersections: Reproductive Genome Editing at a Crossroads', highlighted the need to engage feminist critiques of the fertility industry, particularly in light of proposals to use gene editing as a reproductive technology. Numerous persistent problems have been identified in this profit-driven and poorly regulated global industry, including lack of research into long-term health effects of egg retrieval and the strong potential for exploitation in cross-border commercial surrogacy. Feminist perspectives bring needed attention back to the women and pregnant persons who would bear the risks of experimentation with heritable genome editing (Adams, 2019; Dickenson and Darnovsky, 2019). They would face the burdens and risks of egg retrieval (for research, or for their own or others' use in reproduction), pregnancy and birthing (for their own children or for intended parents), and raising children. They could also face pressures to undergo IVF and other invasive and risky procedures in the absence of infertility, solely to access heritable genome editing.

The panelists discussed how commercialization in the fertility industry might affect the development of heritable genome editing. As panelist, Amrita Pande, warned:

> The way this industry will expand—it has already expanded way beyond the scientific impulse—is what is so dangerous about these new technologies. Even if you want to assume that the scientific impulse was only noble, once it reaches the market it has a life of its own.

Existing reproductive technologies already reflect gender bias and inequality through their use for gender selection, particularly in cultures with strong son preference (Hvistendahl, 2011). Panelist, Kavita Ramdas, asked: 'Why would we not expect those same biases—whether they are pro-whiteness, whether they are pro-male, whether they are pro-lighter skinned, taller, whatever the predilections are—to translate into this new technology?' Pande agreed that these concerns about unequal effects apply not only to future technologies, but also to current practices: 'The desire, the temptation to genetically modify the future generation in a hunt for an assumed perfect race, perfect baby, perfect future generation, is not mere dystopia'.

The discussion also included broader critiques of reproductive oppression and eugenics. The history and ongoing instances of forced or coerced sterilization highlight the heightened risks for women of colour, poor women, women in developing countries, and women with disabilities. Pande rejected the simplistic line that is often drawn to distinguish uses of genetic technologies driven by individual parental decisions from so-called 'old eugenics' – the state-sponsored eugenics programs of the early 20th century. She noted that: 'States, by allowing the fertility industry to operate unhindered, are essentially reaffirming existing structures of inequality and legitimizing new forms of race- and gender-based inequalities' (Ramdas et al, 2023).

Heritable genome editing would also launch in a world where private and state forces have combined to shape an increasingly bifurcated landscape for reproduction. Biotech companies and fertility doctors are racing to offer fine-grained control and unlimited reproductive 'choices' for those who can pay, while increasing restrictions on reproductive rights deprive many women of basic decisions concerning bodily autonomy. The increasing alignment on pronatalist stances among authoritarian governments, far-right political movements, and Silicon Valley tech elites is also concerning (Black, 2022; Shanks, 2023; Munn, 2024; Talbot, 2024).

A broader view on gender justice and heritable genome editing can also illuminate power relations between global North and South. It highlights how global economic inequalities create conditions in which surrogacy offers an attractive alternative to low paying jobs for women in developing countries. Including global South perspectives makes it much more likely that the enormous public health opportunity costs of investing in repro-genetic technologies will be considered.

Taken as a whole, these webinars mark an important effort to make a wide range of 'missing voices' more visible in the ongoing debates about heritable genome editing.

A new framework for considering heritable genome editing

A second project under the umbrella of the Missing Voices Initiative focuses explicitly on the need to establish policies on heritable genome editing that are grounded in equity, social justice, and human rights. The Gender Justice and Disability Rights Coalition of the Missing Voices Initiative is an international coalition of civil society advocates and publicly engaged scholars committed to gender justice and disability rights. The Coalition has produced a groundbreaking set of governance principles for heritable genome editing based on intersectional social justice perspectives (Gender Justice and Disability Rights Coalition, 2024).

The 11 principles, developed to guide policy making on, and public engagement with, heritable genome editing, have an expansive scope that goes far beyond existing frameworks to guide governance of these technologies. They bring to the forefront of the conversation the health, rights, and freedom of women and pregnant persons, as well as the full inclusion and worth of all people with disabilities.

The principles' feminist, anti-eugenic lens illuminates how the harms of heritable genome editing would fall most heavily on groups with the least power and resources. They take an intersectional approach that engages multiple aspects of social justice that are intertwined with gender justice, reproductive justice, and disability rights.

The document concludes that 'there is no persuasive justification for pursuing heritable human genome editing'. It calls on governments and international bodies to prioritize the principles articulated in the document and 'apply the frameworks of gender, disability, racial, reproductive, economic, environmental, and LGBTQ rights and justice, human rights, Indigenous sovereignty, and the rights of children and future generations in all policy concerning heritable genome editing. Our future depends on it' (Gender Justice and Disability Rights Coalition, 2024).

Conclusion

Despite the rapidly proliferating consequences of science and technology on our lives, few avenues exist for people who take social justice seriously to participate in democratic governance of science. This chapter has pointed towards new ways for civil society advocates and publicly engaged scholars committed to social justice and human rights to influence consequential discussions about heritable genome editing. These voices are essential, because we cannot understand how heritable genome editing may affect prospects for equality in the future without considering how present inequalities would shape its use and impacts. As Roberts explained: 'The answers from people who have the greatest stake in radical social change will be very different from those who have a stake in the inequitable and unjust status quo – who are the very people who have been most dominant in these debates about human genome modification' (Roberts et al, 2022).

An expansive conversation grounded in social justice and human rights will nourish richer understandings and encourage wiser policies regarding future developments in human genome editing. These efforts are necessary to ensure that consideration of these powerful technologies is steered by a broad range of human values and a commitment to equity and inclusiveness.

Note

1 Recordings of all events are available from: www.geneticsandsociety.org/internal-content/missing-voices-initiative.

References

Adams, R. (2019) 'Gene-edited babies don't grow in test tubes – mothers' roles shouldn't be erased', *The Conversation*, 29 May, Available from: http://theconversation.com/gene-edited-babies-dont-grow-in-test-tubes-mothers-roles-shouldnt-be-erased-117070 [Accessed 15 October 2023].

Andorno, R., Baylis, F., Darnovsky, M., Dickenson, D., Haker, H., Hasson, K., et al (2020) 'Geneva statement on heritable human genome editing: The need for course correction', *Trends in Biotechnology*, 38(4): 351–354, Available from: https://doi.org/10.1016/j.tibtech.2019.12.022.

Baltimore, D., Baylis, F., Berg, P., Daley, G.Q., Doudna, J.A., Lander, E., et al (2015) 'On human gene editing: International Summit Statement', *News Release*, Washington, DC: National Academies Press (US), Available from: https://www.nationalacademies.org/news/2015/12/on-human-gene-editing-international-summit-statement [Accessed 14 October 2023].

Baylis, F. (2019a) *Altered Inheritance: CRISPR and the Ethics of Human Genome Editing*, Cambridge, MA: Harvard University Press.

Baylis, F. (2019b) 'Human genome editing: Our future belongs to all of us', *Issues in Science and Technology*, 29 April, Available from: https://issues.org/our-future-belongs-to-all-of-us/ [Accessed 23 July 2021].

Baylis, F. and Darnovsky, M. (2019) 'Scientists disagree about the ethics and governance of human germline editing', *The Hastings Center Bioethics Forum Essay*, 17 January, Available from: www.thehastingscenter.org/scientists-disagree-ethics-governance-human-germline-genome-editing/ [Accessed 12 October 2023].

Baylis, F., Darnovsky, M., Hasson, K., and Krahn, T.M. (2020) 'Human germline and heritable genome editing: The global policy landscape', *The CRISPR Journal*, 3(5): 365–377, Available from: https://doi.org/10.1089/crispr.2020.0082.

Baylis, F. and de Vries, J. (2021) 'Equity and access need to be at the forefront of innovation in human genome editing', *The Conversation*, 12 July, Available from: http://theconversation.com/equity-and-access-need-to-be-at-the-forefront-of-innovation-in-human-genome-editing-161794 [Accessed 23 July 2021].

Black, J. (2022) 'Billionaires like Elon Musk want to save civilization by having tons of genetically superior kids. Inside the movement to take "control of human evolution"', *Business Insider*, 17 November, Available from: www.businessinsider.com/pronatalism-elon-musk-simone-malcolm-collins-underpopulation-breeding-tech-2022-11 [Accessed 16 October 2023].

Buchanan, A., Brock, D.W., Daniels, N., and Wikler, D. (2000) *From Chance to Choice: Genetics and Justice*, Cambridge: Cambridge University Press.

Burall, S. (2018) 'Rethink public engagement for gene editing', *Nature*, 555(7697): 438–439, Available from: https://doi.org/10.1038/d41586-018-03269-3.

Burke, T.B. (2017) 'Gene therapy: A threat to the Deaf community?', *Impact Ethics*, 2 March, Available from: https://impactethics.ca/2017/03/02/gene-therapy-a-threat-to-the-deaf-community/ [Accessed 15 October 2023].

Center for Genetics and Society (2023) 'CGS provides comments to National Council on Disability about heritable gene editing', *Biopolitical Times*, 15 May, Available from: www.geneticsandsociety.org/biopolitical-times/cgs-provides-comments-national-council-disability-about-heritable-gene-editing [Accessed 14 January 2025].

Charo, R.A. (2019) 'Rogues and regulation of germline editing', *New England Journal of Medicine*, 380(10): 976–980, Available from: https://doi.org/10.1056/NEJMms1817528.

Cokley, R. (2017) 'Please don't edit me out', *Washington Post*, 10 August, Available from: www.washingtonpost.com/opinions/if-we-start-editing-genes-people-like-me-might-not-exist/2017/08/10/e9adf206-7d27-11e7-a669-b400c5c7e1cc_story.html [Accessed 1 December 2020].

Council of Europe (1997) 'Convention for the protection of human rights and dignity of the human being with regard to the application of biology and medicine: Convention on human rights and biomedicine', Available from: www.coe.int/en/web/conventions/full-list/-/conventions/rms/090000168007cf98 [Accessed 19 September 2020].

Creary, M.S. (2023) 'Racism and the textures of visibility', *American Journal of Bioethics*, 23(7): 109–110, Available from: https://doi.org/10.1080/15265161.2023.2207522.

Darnovsky, M. and Hasson, K. (2024) 'Untangling CRISPR's twisted tales', in N. Baer (ed), *The Promise and Peril of CRISPR*, Baltimore: Johns Hopkins University Press, pp 45–70.

Darnovsky, M., Lowthorp, L., and Hasson, K. (2018) 'Reproductive gene editing imperils universal human rights', *OpenGlobalRights*, 15 February, Available from: www.openglobalrights.org/reproductive-gene-editing-imperils-universal-human-rights/ [Accessed 13 October 2023].

Dickenson, D. and Darnovsky, M. (2019) 'Did a permissive scientific culture encourage the "CRISPR babies" experiment?', *Nature Biotechnology*, 37(4): 355–357, Available from: https://doi.org/10.1038/s41587-019-0077-3.

Dryzek, J.S., Nicol, D., Niemeyer, S., Pemberton, S., Curato, N., Bachtiger, A., et al (2020) 'Global citizen deliberation on genome editing', *Science*, 369(6510): 1435, Available from: https://doi.org/10.1126/science.abb5931.

Dzau, V.J., McNutt, M., and Bai, C. (2018) 'Wake-up call from Hong Kong', *Science*, 362(6420): 1215, Available from: https://doi.org/10.1126/science.aaw3127.

Garland-Thomson, R. (2019) 'Welcoming the unexpected', in E. Parens and J. Johnston (eds), *Human Flourishing in an Age of Gene Editing*, Oxford University Press, pp 15–28.

Garland-Thomson, R. (2020) 'How we got to CRISPR: The dilemma of being human', *Perspectives in Biology and Medicine*, 63(1): 28–43, Available from: https://doi.org/10.1353/pbm.2020.0002.

Gender Justice and Disability Rights Coalition (2024) 'Social justice and human rights principles for global deliberations on heritable human genome editing' Center for Genetics and Society, Available from: www.geneticsandsociety.org/internal-content/principles-heritable-genome-editing.

Generations Ahead (2010) 'A disability rights analysis of genetic technologies: Report on a convening of disability rights leaders' Generations Ahead, Available from: www.redaas.org.ar/archivos-recursos/370-GenerationsAheadDisability.pdf [Accessed 15 October 2023].

Genome editing technologies: some clarifications but no revision of the Oviedo Convention (2021) 'Human rights and biomedicine', Available from: www.coe.int/en/web/bioethics/-/genome-editing-technologies-some-clarifications-but-no-revision-of-the-provisions-of-the-oviedo-convention [Accessed 12 October 2023].

Hvistendahl, M. (2011) *Unnatural Selection: Choosing Boys Over Girls, and the Consequences of a World Full of Men*, 1st edn, New York: Public Affairs.

International Commission on the Clinical Use of Human Germline Genome Editing | National Academies (no date), Available from: www.nationalacademies.org/our-work/international-commission-on-the-clinical-use-of-human-germline-genome-editing#sectionProjectScope [Accessed 12 October 2023].

Jasanoff, S. and Hurlbut, J.B. (2018) 'A global observatory for gene editing', *Nature*, Available from: https://doi.org/10.1038/d41586-018-03270-w.

Kleiderman, E., Ravitsky, V., and Knoppers, B.M. (2019) 'The "serious" factor in germline modification', *Journal of Medical Ethics*, 45(8): 508–513, Available from: https://doi.org/10.1136/medethics-2019-105436.

Lagu, T., Haywood, C., Reimold, K., DeJong, C., Walker Sterling, R., and Iezzoni, L.I. (2022) '"I am not the doctor for you": Physicians' attitudes about caring for people with disabilities', *Health Affairs*, 41(10): 1387–1395, Available from: https://doi.org/10.1377/hlthaff.2022.00475.

Lander, E., Baylis, F., Zhang, F., Charpentier, E., Berg, P., Bourgain, C., et al (2019) 'Adopt a moratorium on heritable genome editing', *Nature*, 567: 165–168, Available from: https://doi.org/10.1038/d41586-019-00726-5.

Macintosh, K.L. (2018) *Enhanced Beings: Human Germline Modification and the Law*, Cambridge University Press.

MacKellar, C. (2021) 'Why human germline genome editing is incompatible with equality in an inclusive society', *The New Bioethics*, 27(1): 19–29, Available from: https://doi.org/10.1080/20502877.2020.1869467.

Meredith, S., Weiss, S., Kleinert, H.L., and Tyrrell, C. (2023) 'The impact of implicit and explicit bias about disabilities on parent experiences and information provided during prenatal screening and testing', *Disability and Health Journal*, p 101514, Available from: https://doi.org/10.1016/j.dhjo.2023.101514.

Mills, P. (2020) 'Heritable human genome editing: the National Academies/Royal Society report', *The Nuffield Council on Bioethics*, 18 October, Available from: www.nuffieldbioethics.org/blog/heritable-human-genome-editing-the-national-academies-royal-society-report [Accessed 14 October 2023].

Mills, P. (2024) 'A therapeutic fallacy', in N. Baer (ed), *The Promise and Peril of CRISPR*, Baltimore: Johns Hopkins University Press, pp 172–184.

Munn, L. (2024) 'Pronatalism is the latest Silicon Valley trend. What is it – and why is it disturbing?' *The Conversation*, Available from: http://theconversation.com/pronatalism-is-the-latest-silicon-valley-trend-what-is-it-and-why-is-it-disturbing-231059 [Accessed 31 October 2024].

National Academies of Sciences, Engineering, and Medicine (2017) *Human Genome Editing: Science, Ethics, and Governance*, Washington, DC: The National Academies Press, Available from: https://doi.org/10.17226/24623.

National Academy of Medicine, National Academy of Sciences, and The Royal Society (2020) *Heritable Human Genome Editing*, Washington, DC: The National Academies Press, Available from: https://doi.org/10.17226/25665.

Ramdas, K., Cattapan, A., Pande, A., and Fogel, S.B. (2023) 'Feminist intersections: Reproductive genome editing at a crossroads [webinar]', *Center for Genetics and Society*, 28 March, Available from: www.geneticsandsociety.org/internal-content/feminist-intersections-reproductive-genome-editing-crossroads.

Roberts, D. (2019) 'Whose conception of human flourishing?', in E. Parens and J. Johnston (eds), *Human Flourishing in an Age of Gene Editing*, Oxford, New York: Oxford University Press, pp 201–211.

Roberts, D., Tsosie, K., Reynolds, M., and Hasson, K. (2022) 'Equity, sovereignty, and racial justice: Beyond access in debates on human genome editing [webinar]', *Center for Genetics and Society*, 14 September, Available from: www.geneticsandsociety.org/internal-content/equity-sovereignty-and-racial-justice-beyond-access-debates-human-genome-editing.

Rulli, T. (2019) 'Reproductive CRISPR does not cure disease', *Bioethics*, 33(9): 1072–1082, Available from: https://doi.org/10.1111/bioe.12663.

Salkever, A. and Wadhwa, V. (2017) 'When baby genes are for sale, the rich will pay', *Fortune*, 23 October, Available from: https://fortune.com/2017/10/23/designer-babies-inequality-crispr-gene-editing/ [Accessed 16 October 2023].

Sarewitz, D. (2015) 'CRISPR: Science can't solve it', *Nature*, 522(7557): 413–414, Available from: https://doi.org/10.1038/522413a.

Shanks, P. (2023) 'Natalism and hipster eugenics', *Biopolitical Times*, 21 September, Available from: www.geneticsandsociety.org/biopolitical-times/natalism-and-hipster-eugenics [Accessed 16 October 2023].

Silver, L.M. (1998) *Remaking Eden*, 1st edn, New York: HarpPeren.

Talbot, M. (2024) 'J. D. Vance and the right's call to have more babies', *The New Yorker*, 5 August, Available from: www.newyorker.com/news/daily-comment/j-d-vance-and-the-rights-call-to-have-more-babies [Accessed 31 October 2024].

Tsaplina, M. and Stramondo, J.A. (2020) '#WeAreEssential: Why disabled people should be appointed to hospital triage committees', *The Hastings Center Bioethics Forum*, 15 May, Available from: www.thehastingscenter.org/weareessential-why-disabled-people-should-be-appointed-to-hospital-triage-committees/ [Accessed 15 October 2023].

Tsosie, K.S. (2021) 'Inclusion without equity: the need to empower Indigenous genomic data sovereignty in precision health', in *Remapping Race in a Global Context*, Routledge, pp 148–154.

Wertz, D.C. and Knoppers, B.M. (2002) 'Serious genetic disorders: Can or should they be defined?', *American Journal of Medical Genetics*, 108(1): 29–35, Available from: https://doi.org/10.1002/ajmg.10212.

WHO Expert Advisory Committee on Developing Global Standards for Governance and Oversight of Human Genome Editing (2021) 'Human genome editing: a framework for governance', Geneva: World Health Organization, Available from: www.who.int/publications-detail-redirect/9789240030060 [Accessed 23 July 2021].

Yee, S., Taylor-Parker, L., Burke, T.B., and Cokley, R. (2022) 'Forging new disability rights narratives about heritable genome editing [webinar]', *Center for Genetics and Society*, 14 November, Available from: www.geneticsandsociety.org/internal-content/forging-new-disability-rights-narratives-about-heritable-genome-editing.

13

Free and Equal: The (Il)Liberal Promise of Human Genome Editing

J. Benjamin Hurlbut

Introduction

Chiselled into the façade of the US Supreme Court building is the phase 'equal justice under law'. Since the 17th century, this principle of equality has increasingly been embedded in conceptions of democracy, justice, and the rule of law.

In the 21st century, equality under law is as fixed an ideal in the democratic imagination as it is in the stone structure of the highest court of the land. The conception of human equality is about as close to a consensus norm as there is amongst contemporary democratic societies. Particularly since the Second World War and the advent of human rights as an international project, equality is taken as a moral fact, a kind of axiom upon which elaborations of human integrity, dignity, and rights are grounded (Moyn, 2010; Waldron, 2017). The idea of equality undergirds the authority and legitimacy of law itself. The concept of equality is an essential element in the normative vocabulary in which political questions of representation, legitimacy, just (and unjust) distribution, and rights are evaluated and contested.

Although equality was a project that modern democracy set for itself at its inception – 'all men are created equal', it required violent transformation and constitutional reform to open the way to the forms of political and legal understanding derived from the phrase on the court's edifice. And these transformations continue; the task of elaborating this concept is ongoing. What promise is held within that capacious and yet ambiguous concept is immensely important not only for its further elaboration and definition, but for the possibilities it holds for the solidification of past achievements, or for their potential attenuation.

This is so especially in arenas of technology. Questions about the social implications of technology in relation to equality tend to focus on the consequences of the use of a technology. Will such and such novel technology produce social conditions that enhance or undermine equality? Do digital technologies that 'democratize' the public sphere contribute to equal opportunity for expression of ideas? Does artificial intelligence, built to produce equal treatment through mechanical objectivity, actually encode inequality in subtle algorithmic biases? Such questions, though important, are typical. Equality and inequality tend to be seen as consequences of technological perturbations to existing distributive arrangements in a society.

Rather less attention is given to the life of the concept itself, particularly where its putative fixity is brought into relation with forces of profound social transformation. Technologies of equality – technologies that promise to move humanity closer to realizing the ideal of equality in practice – have effects not only through the downstream consequences of their use, but also in subtle rearticulation of the concept of equality itself. Any controversy over the goodness or badness – the equality enhancing or inhibiting potential – of these technologies is, therefore, also a controversy over what equality means: over how the ideal is elaborated, understood, and aspired to, and with what consequent reconfigurations and enactments.

Technology is a particularly important cite to see these dynamics in play because ideas and ideals tend to get codified in subtle ways through technological regimes. Aspirations become programmes of innovation and intervention, which in turn become products and uses. Norms that come to be embedded in technological regimes tend to travel with them and assert themselves through uptake and use.

For a technology, such as heritable human genome editing, in which human life itself is the locus of intervention and transformation, this dynamic is particularly significant. Intervening in human life in the name of improving health and wellbeing may entail a (re)conception of what the human is, such that there is a moral warrant to act upon it. The arguments and judgements that initially authorize the use of a technology may become less explicit over time, slipping into unspoken and tacit logics of technological practices even as they thereby become normalized and part of cultural commonsense. Testing for trisomy 21 has followed this path in some societies: prenatal screening has become a normalized and largely unquestioned practice. As testing has been built into the routines of prenatal medical care, so too has the assumption that lives affected by Down's syndrome are undesirable, and that it is a right and responsibility of expecting parents to make an informed choice about whether an affected child should be born. This is but one of many examples of how normative conceptions, that were once front of mind, can slip into subterranean territory, even as the use (and, thus, the acceptance and affirmation) of a technology is often itself taken as evidence of the societal acceptance of this normative transformation.

Equality under law itself offers the remedy to address failures to live up to the ideal: through challenges to, and through, the law, one can bring equality into the open as an essentially contested concept and test the limits of established conceptions. But 'equality under technology' affords no equivalent right. Thus, for a technology, such as heritable human genome editing, in which the concept of equality is itself subject to mutation and transformation, it is imperative to examine how 'equality' is conceptualized, situated in, and refracted through, the practices of technological innovation and ethical sense-making undertaken in its name. This raises questions not only about the consequences of adopting and using heritable genome editing, but about how we approach those questions. What forums will we use? What frameworks of inquiry and debate will guide us? Moreover, we need to consider the foundational norms, such as equality, that are both assumed and challenged by the governance issues surrounding genome editing.

This chapter examines deliberation over heritable genome editing to understand how the idea of equality is situated in relation to conceptions of the human person, to ideas of the warrant for meliorist genetic intervention, and most importantly, of notions of rightly ordered deliberation and governance that have shaped deliberation around heritable genome editing over the last decade.

A technology of inequality?

Most of the discussion of the implications of human genetic engineering for equality have focused on the possibility that genetic technologies, unevenly deployed, will reproduce social and economic inequalities through biological differences: the rich will enhance their children such that they enjoy unequal genetic advantages relative to their un-engineered, poorer peers. This approach treats inequality as a consequence of human genetic engineering – that is, as a state that the technology of genetic engineering will bring about unless measures are taken to limit this risk. Inequality in this scenario is first a biological consequence, that is, divergent genetics, and then a social one, a social structure that is reshaped by the distributive effects of genetic divergence. Some predict a kind of genetic class structure, others even foresee evolutionary divergence and the formation of separate human species. These worries have taken centre stage in discussions about the appropriate and inappropriate applications of human genome editing, and the forms of governance it requires. This one example among many is drawn from the World Economic Forum's 2022 Global Risk Report:

> In the longer run, if (or when) gene-editing technologies become available to enhance physical, cognitive or behavioral capabilities, these could result in a society of genetically enhanced haves and the merely natural have-nots. (World Economic Forum, 2020)

Two decades ago, US geneticist Lee Silver sketched a similar, dystopian, future in which the 'gen rich' – the enhanced – have diverged evolutionarily from the 'gen poor', those who could not (or would not) access the same technologies of self-directed evolution (Silver, 2002). This speciation scenario, while extreme, is not limited to quasi-science, quasi-fiction accounts like Silver's. It is a worry that has figured in some of the most prominent and influential statements about the urgency of adequate governance of genome editing that have been made in recent years (Lander et al, 2019). These accounts point to inequality as a potential downstream consequence of the use of genetic technologies – or, rather, of a distributively unequal use of them.

These are important concerns, and there is good reason to believe that genetic technology would, if made available in the reproductive technology marketplace, follow the same patterns of socio-economic distribution as other reproductive technologies. Indeed, if made available for applications for which there would be a large market, for example, intervening in the genetic risk factors for common conditions like high cholesterol or Alzheimer's risk, these inequalities might be intensified.

The philosopher, Jürgen Habermas, has offered a somewhat different critique of genetic technology. He highlighted the concern that with the advent of technologies that allow for genetic alteration of individuals, there is also a potential for a change in human self-understanding which has the potential to destabilize a shared imagination of human equality.

> Genetic manipulation could change the self-understanding of the species in so fundamental a way that the attack on modern conceptions of law and morality might at the same time affect the inalienable normative foundations of social integration. (Habermas, 2003, p 26)

The collective commitment to equality, which sits at the bedrock of the inalienable normative foundations of social integration is, he argues, a kind of necessary fiction: a norm whose durability depends upon the collective belief that it is true – and that deteriorates when there is a basis for believing that it is not. In other words, for a society that is committed to equality to exist, a sort of *de facto* presumption of equality must exist as a bedrock of that society. And if there is an alienation of that presumption, or even an attenuation or embarrassment through interventions in the human, that fundamentally differentiate human beings from one another, that bedrock may become unstable and fractured. That may happen irrespective of the direct form and material effects intentionally created by genetic difference. In other words, the idea of genetic difference, that direct intentional genetic modification would produce, would itself be the source of this social effect, regardless of the biological effects of the intervention.

> The decoding of the human genome opens up the prospect of interventions that cast a particular light on a condition of our normative self-understanding, a condition that, although natural and thus far unthematized, now turns out nonetheless to be essential. Up to now, both the secular thought of European modernity and religious belief could proceed on the assumption that the genetic endowment of the newborn infant … lay beyond any programing and deliberate manipulation on the part of other persons … [T]his rather ordinary contingency proves to be – in the moment we can master it – a necessary presupposition for being able to be oneself, and for the fundamentally egalitarian nature of our interpersonal relationships. (Habermas, 2003, pp 12–13)

Habermas' argument is subtly but importantly different from the prediction that genetic engineering could create materially divergent groups of people – genetic 'haves and have-nots'. A genetic intervention, that may not make substantive difference in the biological nature of those individuals, could nevertheless engender a change to the self-understandings of people – of their givenness versus their scriptedness and conditionality on predefined genetic parameters – and thus to the understanding of their relations. In this sense, a minor genetic intervention, which is to say one that does not manifest as a significant biological difference, could nevertheless have significant socio-political consequences, since it would represent, on Habermas's account, a kind of repudiation of presumptive human equality.

Habermas's solution to this problem is to avoid severing secular society from important resources of meaning. He points in particular to resources of meaning that reside in religious discourse, which, he suggests, can enter public discourse by 'translating' religious meaning into secular discourse. He finds at the borderlands between religion and secularity, a zone of exchange that requires conveying the conceptual property of religion into public space by shedding its religious trappings.

For present purposes I want to set aside the question of whether Habermas's solution is sufficient, and so too the constructions of 'religious' and 'secular' on which it relies. Suffice it to say that 'the demarcation between religion and the secular is made, not simply found' (Calhoun et al, 2011), and thus the boundary-work that Habermas is engaged in itself requires critical interrogation. But for purposes of this chapter, I want to note simply that the most prominent European theorist of deliberative democracy, in noting the threat of technological capacity to what is the most fundamental 'grundnorm' that undergirds 21st-century constitutional democracy, calls for *a priori* constraints on public ethical deliberation, even as he asserts that its narrowness is the source of the problem. We need, he argues, an expanded repertoire of public moral imagination to preserve the essential commitment to human equality. But that repertoire must confirm to norms of secular public reason: that is, it must check what John Rawls' called 'comprehensive doctrines' at the door and adhere to secular and (putatively) universally intelligible terms (Rawls, 2005). 'Secularity' is made a litmus test for the admissibility of moral discourse in public space.

There is no social institution in political modernity that claims a more central and secure position within the territory of secularity than science. Secularity, however, is not given in advance. It is an enactment, 'made, not simply found'. So too is science, or rather, the public authority of scientific expertise. Expertise is a political achievement in the sense that deference to the public authority of scientists speaking in the idiom of expertise does not come automatically from asymmetrical command of technical knowledge. Achieving a position of expertise requires that the would-be expert, one way or another, wins the power to speak to a public that listens and defers. In this sense, expertise is the achievement of a position of political inequality; the few speak for, and decide for, the many. In idealized form, the power of the few to know (and to decide on the basis of knowledge) on behalf of the many, follows from delegation – from the people surrendering a delimited dimension of their political sovereignty to an expert agent. In practice, however, the dynamics of delegated (and appropriated) expert governance are more complex. Central to them are the bases – the reasons – upon which claims to the authority of science to know authoritatively for society and to configure the parameters as well as the terms of public deliberation and judgement. Habermas' uncritical invocation of the religious/secular boundary begs attention to the way that scientists as standard-bearers of universality and, thus, secularity, assert jurisdiction over delimiting the 'resources of public meaning' which shape and frame public ethical judgement.

Those reasons are essential for examining the implications of a technology, such as genome editing, for equality for two reasons: first, because they shape the conditions under which that concept is subject to transformation, both materially through technological use (that is, the genetic haves and have-nots scenario (Bloomberg, 2024)) and conceptually through mutations of 'our normative self-understanding' (Habermas' worry); and second, because they represent interventions in public discourse rooted in the assertion of legitimately unequal political standing: science can and should speak first and with outsized authority. As I show below, this second form of inequality profoundly shapes the conditions of possibility for democratic reflection – and correction – on the first.

Equality and universality

The project of human genetic manipulation has not entered into social life from nowhere. It was, and is, necessarily embedded in, emergent out of, and unfolding within, a thick socio-technical context. Whether 'improving' the life of the child-to-be, satisfying procreative desires or 'reproductive rights' of the would-be parent(s), or serving a broader social purpose, such as the elimination of genetic disease or enhancement of 'quality birth', the project of developing the technology of heritable genome editing is rooted in ideas of what makes genetic intervention justified and good – of how it serves a worthy and legitimate social purpose. Backing such judgements are norms and processes of deliberation and decisional authority that authorize innovation and the mechanisms of democratic participation (or exclusion) that underwrite them.

We should think of those mechanisms as technologies of legitimation. Political institutions are, fundamentally, about the achievement of legitimacy, whatever the means through which it is achieved. But it is necessary to then ask: What are the institutions in play, and through what modalities do they generate legitimacy? And, as a corollary, what questions do we, the people, need to ask of them to ensure the robustness of those institutions, given that, at the end of the day, they are the custodians of concepts such as equality and the shared political imaginations that underwrite them?

The same questions need to be asked of technologies. Technologies are not merely objects of governance but shape and order social life in profound ways, and are themselves expressions of ideas and ideals about how lives and societies should be ordered and improved. In other words, we cannot (and ought not) to neatly segregate the territories of science and technology on the one hand, and politics on the other.

Thus, the modes of holding political power to account are no less relevant to holding science and technology to account, even if they are underdeveloped with respect to the latter. One symptom of a problematic separation is the way in which highly stylized forms of democratic deliberation are deployed as adjuncts to governance of science and technology. For instance, citizens juries like the UK Citizens' Jury on Genome Editing (2022) are means of producing 'public engagement' in science and technology governance. These are highly choreographed and artificial exercises in democratic deliberation. The choreography is justified by the notion that, for a technical complex domain like genome editing, issues cannot simply be handed to the public because they do not have the requisite knowledge to understand and rationally engage with them. The citizen's jury functions as a kind of stand in technology of deliberation in which a curated 'minipublic' can stand in for the larger society (Fung, 2007). This group is first 'educated' in what the organizers deem to be the knowledge necessary to deliberate informedly, and likewise, is provided with the questions and framings that rightly orient deliberation. Needless to say, this parameterization of deliberation is an exercise of power, though one that justifies itself as a politically neutral expression of expertise. However, the important point is not merely that the deliberation is configured. That is self-evident. The important point is that the deliberation is configured in the name of achieving political equality – that is, overcoming knowledge deficits that cause people to be excluded from reasoned democratic deliberation, and thereby producing free and equal citizen participation in deliberation. The citizens' jury is supposed to be a technology of equality.

This construction of equality authorizes the (unequal) role of experts in setting and guiding deliberative agendas. It also authorizes their role in interpreting the quality and results of a deliberative process, translating the attitudes and utterances of the citizen jurors into conclusions about 'what the public thinks' that can inform – and legitimate – policy making and governance approaches. For instance, at the conclusion of the UK Citizens' Jury on Genome Editing (2022), certain scientific and policy experts in the field were called upon to evaluate the value and relevance of the deliberation for policy making. The experts freely evaluate the actual deliberative exercise in relation to their own imaginations of what good public deliberation should look like (matters that are, be it noted, well outside their professional expertise). 'That jurors seem to have kept an open mind and understood the question they had to answer, instead of just announcing their own perspectives, adds weight to the recommendations of the jury for the experts we spoke to' (Curato et al, 2023, p 23).

The key point is that, with this particular technology of legitimation, the citizen's jury is organized around *a priori* conceptions of what needs to be known about genome editing, what is at stake socially and ethically in its use, and no less importantly, what counts as legitimate – informed, 'open-minded', free and equal – deliberative democratic participation. It is an idealized and stylized construct of political equality that simultaneously authorizes constitutive forms of political inequality. Specifically, it empowers experts to set and shape deliberative agendas and to evaluate the significance of deliberative outcomes afterward. This framework serves as a standard for distinguishing between 'good' deliberation and the uncontrolled, supposedly 'bad' forms that exist in the public sphere.

But this is democratic deliberation only if one understands what is upstream and downstream of it – formation of knowledge, technology and expertise, and governance, respectively – as external and non-political. This is a deeply defective conception of both technology and politics that also underwrites fundamental political inequalities in the authority of science to shape and govern technological futures. The notion that one-shot technologies of democracy, such as citizen juries, satisfy the requirements of democratic participation is itself a symptom of the urgent need for a fundamental reorientation in governance of science and technology.

In the next sections I explore the canonical moves made by scientific experts to assert jurisdiction over deliberative agendas. I show how these moves have directly contributed to the production of heritable genome editing technology itself, as well as to the corollary imagination of genome editing technology as a corrective to inequality.

Expert jurisdictions

In his remarks opening the first International Summit on Human Gene Editing in 2015, Nobel laureate and former president of Caltech David Baltimore said, 'the unthinkable has become conceivable. Now we must face the questions that arise. How, if at all, do *we as a society* want to use this capability?' (Baltimore, 2015).

Baltimore is invoking a familiar narrative: advances in science and technology raise ethical questions. Scientific knowledge leads to technological capacity which, when deployed into society, has social consequences that may be ethically concerning. From science, to technology, to society. This is the so-called linear model of innovation. It is an old (and thoroughly discredited) story of innovation. But what I want to draw attention to is that it is also a story of how and when questions of the good come into play in relation to science and technology. Ethics only enters at the moment of transition from technology to society, in the moment when knowledge and technological capacity have already been produced and all that remains is to deploy them. This is an account of socio-technical change that is widely taken for granted. It presumes relationships between science and public ethical debate, wherein science acts first and society reacts only after scientific and technological advances have been made. This linear narrative both asserts and naturalizes these relations, in effect presenting them as a norm. Society *should* only react after the fact. This positions scientists and technologists as first movers – as acting upstream and independently of social judgements.

Baltimore continued:

> We are taking on a heavy responsibility for our society, because we understand that we could be on the cusp of a new era of human history. The actions we take now will guide us into the future. We are initiating a process of taking responsibility for technology with far reaching implications. (Baltimore, 2015)

I want to draw attention to his use of the first-person plural and the slippage from 'How will *we as a society…*' to '*… we are taking on for our society.*' That slippage is an act of self-empowerment of the few to know, think, and speak for the many. This is appropriation cast as 'taking responsibility'. And this is the logic that it follows: science supplies the common ground as the basis for reasoned public deliberation about matters of science and technology. After all, we need to know what we are talking about before we can talk about it. And therefore, that deliberation must be initiated, led, shaped and, in effect, governed by those who know best: the scientific experts.

In practice, deliberations like the 2015 Summit follow this script. Scientific experts are invited to lay down what is held in common – that is, the repertoire of 'facts' that are the universal priors to value-based deliberation. First comes the panel of scientists to lay out what is known – what is technologically possible, and what is unrealistic and therefore outside the scope of deliberation. They separate 'science fact' from 'science fiction'. I want to draw attention to how this is presumed to be a technology of deliberation and, as such, of equality, but one grounded in a fundamental asymmetry. Shared facts provide a necessary common foundation as well as shared vocabulary. Free and equal participation from this perspective is rooted in a shared epistemic common ground.

But there is another essential logic in play. The trajectory from science to technology and from technology to society is also construed as a trajectory from universality to particularity – from placeless, objective knowledge to historically, culturally and geographically situated societies and subjects; from univocality of knowledge to pluralism of values; from authoritative consensus to individual preferences, from science to merely interested stakeholders.

The narrative of linearity highlights both the connections and separations between universal science – which asserts authority over objective facts – and the distinctiveness and unpredictability of politics and culture. From this perspective, scientific knowledge and technological advancements transcend the contexts of place and culture.

The idea of linearity underwrites three corollary commonplaces, of which I offer only brief sketches here. First is the notion that society necessarily and inevitably lags behind science and technology. In the words of Marcia McNutt, the president of the US National Academy of Sciences, '… as is always the case, the speed at which the science is advancing outpaces society's ability to grasp its implications' (NASEM, 2017b). This is the so-called 'law lag', which some legal scholars have called the 'pacing problem' (Marchant et al, 2011). Suffice it to note that this is a flawed construction. Science and technology are coproduced with social, legal and normative orders. However, I want to specifically highlight how this narrative characterizes law and social norms as necessarily following after, and subject to, transformation by technological developments. Thus, on this narrative, science and technology unfold in as-yet ungoverned territory.

Second, linearity underwrites the notion of technological inevitability. The narrative is as follows: Science inevitably generates technological innovation. Once capacity exists, the story goes, it is bound to be used. A 2015 Nature Biotechnology interview of leading figures in genome editing science and bioethics asked: 'To what extent do you think germline engineering is inevitable?' Most answered that it was (Bosley et al, 2015).

In November 2018, the Chinese scientist He Jiankui, who produced the 'Chinese CRISPR babies', wrote to Steven Quake, his postdoctoral supervisor at Stanford University. Quake immediately sent a text to someone he described as 'an extremely prominent scientist in the field'. The text said: 'First genome edited baby—Done! (In China, where else?)'. The reaction: 'not terribly surprised. Not an oh-my-God thing, like, it's inevitable, like the inevitable has happened' (Belluck, 2019).

The discourse of inevitability, which is widespread in science and technology, has been particularly present around heritable genome editing. For those who predicted it, the notion that human germline editing was inevitable was born out by the babies in China. They invoked that event to argue that limits or prohibitions are not strong enough to hold back the forward march of science and technology, and therefore society must accept and embrace such technologies. In other words, society must accept their inevitability and 'catch up'. A year after the CRISPR babies revelation, Jennifer Doudna wrote that categorical prohibitions, such as moratoria, are 'no longer strong enough countermeasures, and instead stakeholders must engage in thoughtfully crafting regulations of the technology without stifling it' (Doudna, 2019, p 777). Indeed, within days of He's revelation, the leaders of the US National Academies had already declared that, because He had crossed this line, this made it urgent for scientists to define a 'responsible translational pathway' for bringing heritable genome editing into clinical use. The logic was clear: why ask whether it can be used if it is inevitable that it will be? The task of governance requires accepting that use is inevitable and doing what one can to put guardrails around it (Dzau et al, 2018).

And yet inevitability is a self-fulfilling prophecy. He's experiment was not a confirmation of the prediction that heritable genome was inevitable, but a product of it. In a series of extended interviews, which I conducted with him in late 2018 and early 2019, before he was taken into custody by Chinese authorities, He explained that he felt empowered to take this radical step for two primary reasons: First, because someone, somewhere, would be the first to make genetically engineered children. It was inevitable. Here he was simply parroting what many prominent voices in the scientific community took as a self-evident truth. Given this, he asked, why should it not be him? He was simply doing something that would inevitably be done by someone. By acting first, he could, he believed, also take responsibility for doing it well. Thus, acting in light of the inevitable was not, he thought, an act of transgression, but an act of responsibility. The second reason is because science cannot and should not wait for society's consent because it is impossible for society to grasp where science and technology are leading. After all, society and its rules and norms necessarily live in the past – they lag behind. It is incumbent upon the pioneering scientist to take the controversial step so that society can catch up with, and accept, the (inevitable) technological future. He Jiankui said: 'If we are waiting for society to reach a consensus … it's never going to happen …. But once one or a couple of scientists make first kid, its safe, healthy, then the entire society including science, ethics, law, will be accelerated'.

He understood himself to be simply following and playing his part in contributing to the natural trajectory of linear scientific progress. He followed the lead of the leading figures in the scientific community in invoking notions of linearity, lag and inevitability to assert jurisdiction as well as responsibility and authorize action, in effect asserting an intrinsic and natural inequality between science and society in the power to make the future.

Technologies of equality

I have shown how the imagination of linearity, that dominates in scientific circles, leads scientists to arrogate authority to shape the technological future. In the case of heritable genome editing, that arrogation extends to conceptions of human equality itself.

When the CRISPR babies' story broke, the leading scientific voices roundly condemned He. One of the claims that dominated was that the edit he had made was 'medically unnecessary'. He had attempted to inactivate a gene that codes for a T-cell receptor that functions as a gateway for HIV infection. In theory, the effect of the edit would render the edited child immune to HIV. To qualify for the trial, the male member of the married couple had to be HIV-positive and on antiretrovirals. Critics observed that for a child in such a family there is no higher than average risk of contracting HIV.

The notion that it was medically unnecessary has, as its corollary, the notion that there are edits that are medically necessary, and that it is the prerogative of scientific experts to determine them. Yet, as the UK Nuffield Council (2018) observed, germline editing is not a therapeutic intervention in a patient in need of treatment because the patient does not exist. The 'patient', that is, the embryo, is brought into being at the same moment as the 'treatment' is administered. In this sense, germline genome editing is never therapeutic, but is a reproductive technology whose purpose is to fulfil a reproductive desire, such as to have a child free of genetic disease, or immune to HIV. This is illustrated in the discourse that already surrounds preimplantation and prenatal genetic screening, and which is still more pronounced around genome editing: ensuring that the resulting child will be 'unburdened by disability', will have a 'better chance at a full life', 'the best opportunity', 'evening the playing field'. As scholars of disability have long shown, these sorts of evaluations of ability and disability occlude the social conditions that render a difference into a disability. The case of the CRISPR babies is a case in point.

In Chinese society, being HIV positive carries enormous stigma. Indeed, one of the reasons He was easily able to recruit participants in his trial was that IVF clinics will not provide fertility services to individuals who are HIV positive. Although the HIV positive individual may be largely biologically unaffected by infection, it can nevertheless be a profound disability if one's status is known. In attempting to 'genetical vaccinate' the embryos of these couples, He was inoculating them against social stigma as well as the virus itself. It was a genetic solution to a problem of social discrimination and inequality. The participants themselves attest to as much. In a letter written by one of the participating couples to the judges overseeing He's criminal prosecution, they defended He and denied that he had coerced them into participating. Rather, they said, they were driven to participate by the position of marginality and risk they experienced as Chinese citizens marked by HIV.

> As an AIDS sufferer and family member, we firmly and deeply know that it is possible to use a preventative drug to have a healthy child...That drug can cure disease, but it cannot cure prejudice...our participation in the experiment was indeed forced, but we weren't coerced by any person in particular. We were coerced by society.

This poignant statement illustrates both the true nature of the genetic intervention and the vulnerability of those in positions of marginality to the draw of a genetic 'solution' to situations of inequality. Placed in social context, He's intervention is not so easily dismissed as 'medically unnecessary'. At the same time, it reinforces a social situation that it ostensibly seeks to remedy: it sees a socially-circumstantial liability as a disability, and the disability as a genetic disease reparable under the genetic surgeon's molecular scalpel. It is not hard to imagine that there are other genetic conditions that might be seriously disabling in particular social contexts, and genetic modifications that would be seen as similarly emancipating, for instance alteration of skin colour, or other genetically simple elements of physical appearance, that may give rise to discrimination. The individual motivation for seeing such an intervention for one's future child is understandable, even as it is abhorrent. But it is not difficult to see how this sort of genetic intervention, made in the name of the social wellbeing of the future child, lends itself to justification in terms of fairness and equality. Indeed, this was precisely how He explained it to me in a January 2019 interview. The mutation he was attempting to create is not uncommon within certain populations of northern European origin. Why, he asked, should this genetic advantage be limited to them? As he saw it, he was simply using genome editing to equalize the inequalities of the genetic lottery, bringing a beneficial variant from one population into another – one that needed it especially badly. It was genetic engineering in the name of human equality; CRISPR-driven genetic distributive justice.

Some of those who condemned He's experiment as medically unnecessary were also authors of a US National Academies of Sciences, Engineering and Medicine (NASEM) report that counselled deference to culturally situated judgments of what is 'medically necessarily'. The report suggests that there are some cases in which human germline genome editing could be permissible, but these must be restricted to 'preventing a serious disease or condition.' It goes on to explain:

> It is important to note that such concepts as "reasonable alternatives" and "serious disease or condition" embedded in these criteria are necessarily vague. Different societies will interpret these concepts in the context of their diverse historical, cultural, and social characteristics. (NASEM, 2017a, p 8)

I want to draw attention to the interplay of (scientific and technological) universality with the authorization to calibrate to sociocultural particularity. He did precisely what the sentence above describes. He offered (what he presented as) a Chinese assessment of the medical warrant and social acceptability of the genetic intervention, combined with scientific self-authorization and self-governance rooted in the notion of linearity. In other words, following the lead of scientific authorities, such as the authors of the NASEM report, he asserted the unequal role of science and scientist to enhance future human life, including by rectifying conditions of genetic inequality.

The He case is illustrative of patterns that extend well behind his particular project to the culture and self-imagination of science that underwrites the arrogation of authority. But it applies also to ideas about what is due to human beings – what makes for good lives, and what forms of equality of opportunity to access the good life, human genetic engineering affords.

The focus of the debate about good and bad life – about 'what sort of people there should be?' – has tended to focus on the children these technologies would produce. This is, without question, an important focus. But as a result, they have tended to overlook the lives that are no less a target of transformation: those who are emancipated from the constraints of natural life's limitations in the project of procreation.

In his opening remarks at the 2022 annual congress of the American Society for Reproductive Medicine (ASRM), Society President and Yale School of Medicine Professor Hugh S. Taylor described a future in which 'all reproduction takes place outside of the human body'. The theme of the meeting was 'reproduction reimagined'. It looked back on 40 years of IVF in the United States, and forward to the next 40. Taylor's remarks introduced the first keynote speaker of the meeting, Jennifer Doudna, the Nobel Laureate inventor of CRISPR-cas9 genome editing. Doudna gave a keynote entitled 'CRISPR: Rewriting DNA and the Future of Humanity'. For the community of professionals to whom these remarks were delivered, technologies (such as genome editing) are not merely a means to produce 'better' children. They are a means through which to provide options to their customers – would-be parents. They are the tool kit through which those people are set free to reproduce as they wish, emancipated from the limitations of their own bodies, or from the burden of carrying a pregnancy, or the risk of having a genetically disordered child, or (with genome editing) from the limited assortment of genomes which the intermixing of fixed genetic inheritances can produce. These technologies are offered not merely as instrumental means, but as vehicles of emancipation, the pathway to exercising reproductive rights – including the 'right' to bend the limitations of natural procreation to one's will. There is a concept of liberal individualism built into this technological regime, and with it a notion of equality. This is not equality as a natural fact of human life, but rather equality as a promise offered by technologies that can neutralize differences, equalize heritable advantages, and liberate people from the burden of having children that they believe should not exist. These technologies provide the promise that people can fulfil their imagined right of choice by having the children they want – children who are the kinds of people who they believe should exist.

Conclusion

This chapter has followed ideas of equality through a series of formations that produce inequality – in political voice, in future-making agency, and in moral imagination. I have argued that if we focus too narrowly on the risk of inequality as a consequence of the application of a technology, such as heritable genome editing, we miss the extent to which the very formation and deployment of that technology is already reshaping our understanding of equality. This process influences not only shared understandings of what forms of equality we should strive for but also which existing inequalities we need to address and rectify.

In contemporary democracies, we tend to treat law as the pathway for achieving equality, even if that pathway is recognized as being imperfect. Yet the regimes of technology that order lives and societies are no less implicated than the law in shaping conditions of equality or inequality, and thus should be no less susceptible to scrutiny and democratic challenge. Technology is central to forms of social transformation that have the potential to solidify or attenuate equality, particularly a technology poised to alter the very stuff of human life.

In 21st-century political discourse, the equality of persons is presumed as a kind of political fact: a given that is beyond contestation and whose acceptance is widely seen as a precondition for participation in political life. This way of thinking about equality as a fundamental principle posits it as unchangeable. Persons may be treated unequally, but such treatment is a violation of equality, not a problematization of the ideal as such. Much of the political energy dedicated to the elaboration of equality, therefore, takes the form of seeking to decern, expose, and rectify those social conditions that engender unequal treatment.

However, this chapter has sought to draw attention to the ways ideas of equality are, themselves, at stake in visions of technological futures, and in the logic of authority and authorization that drive them. It has shown how notions of fundamental inequalities of knowledge empower a select few individuals to configure the parameters of public reasoning. This includes defining the conditions for political participation, establishing rules of engagement, and determining the terms of debate. (Un)equal political participation, (un)equal agency, (un)equal freedoms of self-expression and fulfillment of reproductive desire, and imaginations of (un)equal lives are part and parcel of the project of heritable human genome editing. They are reflected in the surrounding political debates, but ultimately they come to shape and be embedded into the very essence of techniques and technologies, as well as the contexts in which they are deployed.

The idea of equality asserts a truth about human lives that can be true only to the degree that societies aspire to live by it. We must attend, therefore, to the social life that the idea of equality takes on in the context of the technologies deployed in its name. We attune ourselves to the ambitions to edit not only the molecular stuff of life, but the moral imagination.

References

Baltimore, D. (2015) 'Human gene editing summit: Context for the summit', Washington, DC, 1 December 2015, Available from: https://vimeo.com/album/3703972/video/149179797

Belluck, P. (2019) 'Gene-edited babies: What a Chinese scientist told an American mentor', *The New York Times*, 14 April 2019, sec. Health, Available from: www.nytimes.com/2019/04/14/health/gene-editing-babies.html

Bloomberg.Com. (2024) 'Genetic haves and have nots?', 5 November 2024, Available from: www.bloomberg.com/news/articles/1999–07-12/genetic-haves-and-have-nots

Bosley, K., Botchan, M., Bredenoord, A., et al (2015) 'CRISPR germline engineering – the community speaks', *Nature Biotechnology*, 33(5): 478–486.

Calhoun, C., Juergensmeyer, M., and VanAntwerpen, J. (eds) (2011) *Rethinking Secularism*, Oxford, NY: Oxford University Press.

Curato, N., Parry, L., and van Dijk, L. (2023) 'Evaluation Report: UK Citizens Jury on Human Embryo Editing', Available from: www.wellcomeconnectingscience.org/wp-content/uploads/2023/07/Evaluation-Report-UK-Citizens-Jury-on-Human-Embryo-Editing.pdf

Doudna, J. (2019) 'CRISPR's unwanted anniversary', *Science*, 366(6467): 777.

Dzau, V.J., McNutt, M., and Bai, C. (2018) 'Wake-up call from Hong Kong', *Science*, 362(6420): 1215.

Fung, A. (2007) 'Minipublics: Deliberative Designs and Their Consequences', in S.W. Rosenberg (ed), *Deliberation, Participation and Democracy*, London: Palgrave Macmillan UK, pp 159–183.

Habermas, J. (2003) *The Future of Human Nature*, Cambridge, UK: Polity.

Lander, Eric S., Baylis, F., Zhang, F., et al (2019) 'Adopt a moratorium on heritable genome editing', *Nature*, 567(7747): 165–168.

Marchant, G.E., Allenby, B.R., and Herkert, J.R. (2011) *The Growing Gap Between Emerging Technologies and Legal-Ethical Oversight: The Pacing Problem*, Springer Science & Business Media.

Moyn, S. (2010) *The Last Utopia: Human Rights in History*, Harvard University Press.

Rawls, J. (2005) *Political Liberalism*, 2nd edn, Columbia University Press.

Silver, L.M. (2002) *Remaking Eden: How Genetic Engineering and Cloning Will Transform the American Family*, New York: Perennial.

UK Citizens' Jury on Genome Editing (1 August 2022) 'Convened by Wellcome Trust', Available from: www.wellcomeconnectingscience.org/project/uk-citizens-jury-on-genome-editing/ [Accessed 16 January 2025].

UK Nuffield Council on Bioethics (2018) *Genome Editing and Human Reproduction: Social and Ethical Issues*, London, UK: Nuffield Council on Bioethics.

US National Academies of Sciences, Engineering and Medicine (NASEM) (2017a) *Human Genome Editing: Science, Ethics, and Governance*, Washington, DC: The National Academies Press.

US National Academies of Sciences, Engineering and Medicine (NASEM) (2017b) *Public Report Release Event for Human Genome Editing: Science, Ethics, and Governance*, Washington, DC: National Academies. https://vimeo.com/205455452.

Waldron, J. (2017) *One Another's Equals: The Basis of Human Equality*, Cambridge, MA; London, England: The Belknap Press of Harvard University Press.

World Economic Forum (2020) 'Global Risks Report 2020', Insight Report, Available from: www.marshmclennan.com/web-assets/insights/publications/2020/january/WEF%20-%20Global%20Risks%20Report%202020%20-%20full%20-%20web.pdf.

Conclusion

Calum MacKellar, Roberto Andorno, and Matthew James

How equality became important

The previous chapters in this volume have carefully examined a number of different angles relating to the question of whether heritable genome editing may be compatible with equality in society. It began by studying the origins of the concept suggesting that equality, in antiquity, was considered important by Stoic philosophers who believed that every rational human individual was deserving of the same dignity and respect. This was because human beings shared the ability to reason which was seen as an aspect of the divine. The New Zealand legal philosopher, Jeremy Waldron, explained: '[T]he Stoics imagined that we cannot commune with the gods unless in some degree we, like they, are rational beings' (Waldron, 2017, p 140). Similarly, the Roman statesman and Stoic scholar Cicero (106 BC–43 BC) indicated:

> This animal – provident, perceptive, versatile, sharp, capable of memory, and filled with reasons and judgement – which we call a human being, was endowed by the supreme god with a grand status at the time of its creation. It alone of all types and varieties of animate creatures has a share in reason and thought, which all the others lack. What is there not just in humans, but in all heaven and earth, more divine than reason? (Cicero, 1999, p 113)

However, Matthew James indicates in his chapter that the concept of equality, in history, was not only based on philosophical arguments but on the religious belief that all human beings were created in the image of God, such as in Judaism and Christianity. He notes that, obviously, equality does not mean that all human beings are identical in terms of capacities or physical traits, but that they possess the same inherent worth. In other words, human beings have some basic qualities common to all but which are not exactly the same. James then explains that the modern notion of equality was formally accepted in the UN Declaration of Human Rights of 1948 as a global legal and political standard.

It is also suggested that the concept of shared equality may arise from society's collective decision to treat individuals equally, often enshrined in the National Constitutions of countries. But as Roberto Andorno points out in his chapter, the contractarian theory is insufficient and has a significant flaw; it lacks a deeper moral justification for equal treatment and remains at a purely formal level. He concludes that only the concept of inherent human dignity, now firmly established in international law, provides a solid moral foundation for equality in society. However, the real meaning of this inherent human dignity may remain unclear to many among the general public.

Discussing equality in the context of heritable genome editing

In the context of discussing the different possible risks to equality of heritable genome editing, it is also important that all sections of society are involved. Indeed, it is unfortunate that the scientists, who dominated the influential national and international committees on the topic seemed to have increasingly sidelined public engagement and influence on the final decisions. In addition, by focussing on biomedical promises, rather than social issues, control of the conversation was handed to the biomedical scientists, physicians, and medical ethicists. No careful consideration was developed in relation to the commercial dynamics and pressures as well as to the social repercussions arising in society as a result of possible systemic inequalities and discrimination (Darnovsky and Hasson, 2024, p 60). As Marcy Darnovsky and Katie Hasson indicate: 'The [US] science academies' reports assume that Heritable Genome Editing will soon be shown safe and effective; they proceed from there to introduce guidelines for how it should be put into practice. Sorting out the question of whether it should be used at all is left in the dust' (Darnovsky and Hasson, 2024, p 60).[1] In this regard, the 2025 Global Observatory for Genome Editing Summit Statement indicated:

> [T]he acceptability of heritable genome editing is not primarily a technical question …, but first and foremost a question about human relations and mutual obligations – parent to child, medicine to patients, society to its members, states to their citizens, and the global community to the common heritage of humankind. (Global Observatory for Genome Editing, 2025)

Because questions relating to the values of future societies are complex, it also means that they require an appropriate amount of time for reflection. This, in turn, often requires supportive resources which can be a serious challenge to those who believe they can quickly address the science and the ethics (Baylis, 2019, p 145). Thus, it is important to create new ways for civil society advocates committed to social justice and human rights to influence consequential discussions about heritable genome editing. Their voices are essential because it is impossible to consider, for example, how such a procedure may affect equality in the future without understanding how present inequalities are influencing society.[2]

The general public and equality in decision making

Interestingly, what may be understood by equality in public participation, even when discussing the very topic of equality, may vary quite significantly in different contexts. It is important, therefore, that the different sections of society are able to discuss and understand the concept of equality in a manner which respects all the different voices.

Accordingly, society should be able to understand the possible risks to equality which may result from any societal acceptance to select in reproductive procedures, including with heritable genome editing. This is because it may undermine the very concept of *inherent and equal value of all persons in society*, meaning that such decisions may change the ethical core of civilized society.[3] In addition, as Hofmann warns: '[S]ocial practices frame and form our social norms and values and that the sum of individual legitimate actions may lead to a society we do not want' (Hofmann, 2017, p 512). The general public must become aware, therefore, that within a society where inequality is rampant, heritable genome editing may just reinforce the advantages of those at the top of the social hierarchy – even if it were made widely accessible.[4]

Because of this, it may be useful to now examine some of the risks to equality arising from heritable genome editing.

Selection risks and the prospective parent(s)

Many future parents may be interested in some form of *selection* through heritable genome editing, in certain circumstances, if ever this procedure becomes feasible. This may be for their own sake in avoiding the generation of a child with genetic disorders. In other words, this may be similar to prenatal or preimplantation selection, which are often considered from the perspective of the interests of prospective parents (Scott, 2007, p 68). However, the decision may not be that simple. Søren Holm explains:

> [A]ny change in reproduction which entails a change in the timing or manner of conception leads to the production of different children (children with different identities). … What we have to decide is whether parents are allowed to choose to bring into the world a child

> with an impairment, when they could have produced another and 'better' child, or whether they are only allowed to bring 'the best possible' child into the world. (Holm, 2000, p 187)

This question is important because the real incentive behind heritable genome editing may be to enable a very small number of individuals to have genetically related children, but for whom the selection of a healthy embryo through procedures, such as preimplantation genetic diagnosis, cannot be considered. In this respect, while it is normal that compassion is shown toward prospective parents who value full genetic kinship, this sentiment can be regarded as a personal desire rather than a medical requirement. This means that it is necessary to weigh the personal desire of a few against the risks arising from heritable genome editing for the whole of society (Darnovsky and Hasson, 2024, p 49).

In addition, while it may be possible to accept that prospective parents may be able to distinguish their attitude towards a child with physical or mental disability from their fears about being unable to cope, this may not always be the case. It may be possible to envisage situations in which the parents' internalisation of negative social constructions of disability would make it almost impossible for them to disentangle their own attitudes concerning disability from their concerns about providing adequate care. Moreover, even if such a distinction could be made and selection, such as with heritable genome editing, could be considered solely because parenting a child with a disability would present a major practical and financial burden and not represent any hostility towards disabled people, the implied message may still be the same (Song, 2007, p 239). In this connection, the UK theological ethicist, Robert Song, argues (in the specific case of selective terminations) that the message would be 'that disabled people are an inconvenient nuisance and allowed into society only on sufferance'. He adds: 'In a society where disabled people still feel tolerated rather [than] actively welcomed and affirmed, abortion connected with disability will continue to be experienced as a sign of rejection' (Song, 2007, p 239).

Selection risks and the possible future child

Prospective parents and society may also want to consider selective procedures, such as heritable genome editing, for the possible future child's sake. This may happen when a risk exists that he or she may be affected by a very serious disorder. The American ethicist Ronald Green indicates: 'Parents have a prima facie obligation not to bring a child into being deliberately or negligently with a health status likely to result in significant greater disability or suffering, or significantly reduced life options relative to the other children with whom he/she will grow up' (Green, 1997, p 10). The UK academic commentator

Simon Reader also states: 'I will take it as uncontroversial that there are harmful abnormalities which spell such suffering for a future child that exercising forms of selective reproduction can be the only humane thing to do' (Reader, 2017, p 9). Similarly, British philosopher David Archard indicates that children should have a reasonable prospect of minimally decent lives adding:

> I argue that it is wrong deliberately to bring into existence an individual whose life we can reasonably expect will be of very poor quality. The individual's life would on balance be worth living but would nevertheless fall below a certain threshold. (Archard, 2004, p 403)

But is this right? Can the eventual suffering of a child ever undermine the inherent value and worth of a life in a civilized society? In her chapter, Pia Matthews explains that the concept of a 'life unworthy of life' questions the moral status of people with severe genetic disability and whether they deserve to exist. Indeed, if the life of a possible future person is unworthy of life, then the non-existence of this person may seem preferable. She also indicates that, by analysing the underlying assumptions and attitudes of those who believe that a life is unworthy of life in the case of severe disability, questions may be asked about how they understand what it means to be a person with inherent value and worth as well as the concept of equality.

Matthews further notes that suggesting that disability can be eradicated ignores the very fact that disability does not exist without someone (a person) living with that disability. Sandy Sufian and Rosemarie Garland-Thomson explain, in their chapter, that their genetic conditions cannot simply be taken away from them as if they were some kind of misspelled word or an awkward sentence in a document. Instead, they describe how they are whole beings, with their genetic conditions forming a fundamental part of who they are. They are also concerned about the mistaken assumption made that people with genetic challenges are in a constant state of suffering, which limits their lives.

Selective risks and existing individuals

It is worth remembering that when any specific selective decision is made, this must be grounded on considerations relating to embodied persons who already actually exist if this decision is to make sense in any rational way.[5] For example, Hofmann writes: '[P]eople do not select foetuses based on chromosomes or other biological traits, but based on characteristics of living persons with specific disabilities' (Hofmann, 2017, p 514).

Moreover, because the decision to select through heritable genome editing is one which would generally become public, and may even become publicly acceptable, this may have detrimental consequences to persons who already exist with the unwanted traits. For example, persons with a disability may be seriously challenged with their self-perception as well as the way they value their own lives (German National Ethics Council, 2003, p 86). In other words, the core problem with deselective procedures is the message being received by a number of existing individuals, including some with disability.

In this respect, Sufian and Garland-Thomson are concerned that heritable genome editing will, in the future, ensure that persons like themselves will be deselected out of existence without others even noticing. They even suggest that such a version of a future, where people like them are selected out, limits their ability to live in the present. In this way, the academic commentator on disability, John Swinton, in Scotland indicated: 'To pretend that they [someone with a disability] are welcome and then to engage in social practices which are designed to ensure that their existence 'won't happen again' is loveless and inhospitable' (Swinton, 2007, p 19).

Thus, if society genuinely believes that every person has absolutely equal value and worth, then it must also, to remain rational, believe that every possible future person holds the same value and worth in a similar way.

Selection risks and ableism

Another risk raised by heritable gene editing concerns the problem of 'ableism'; a term which was coined in the 1970s to express the negative experiences of persons who cannot fulfil species-typical physical, mental, neurological, or cognitive ability expectations (Wolbring, 2012). The concept includes a number of beliefs, processes, presumptions, and practices, which uphold the normal human standard while devaluing those who cannot fulfil these characteristics. This means that an ableist perspective implies that persons without a disability are the norm in society, and that those who are affected by disabilities must endeavour to become part of the norm if they do not want to be marginalized. In this way, the concept of ableism seeks to express similar concerns to those who experience other forms of discrimination, such as racism, sexism, and ageism. For this reason, Matthews explains that, if it is accepted that some lives are not worth living, then this buys into ableism and rejects the notion of the equal moral worth of all human beings.

Interestingly, it seems that an element of ableism discrimination may exist in the anti-discrimination agendas themselves (a form of 'discriminationism') with some forms of anti-discrimination programmes being seen as more acceptable than others. For example, in 2001, the British journalist Polly Toynbee denounced the UK disability rights commission for seeking similar objectives to other anti-discriminatory groups:

> In recent years the militant disability movement has developed a whole new confident and strident identity – no more pathetic cripples pleading in soppy charity ads, but a strong demand for equal rights in the tradition

> of the battles for black, women's and gay rights … After women's pride, black pride and gay pride, proud-to-be-disabled seems to make sense too. But each of these four campaigns is different. Each group's identity, cohesiveness and claim to victimhood is based on markedly different circumstance. Disability campaigners have over-identified with other civil rights issues, talking as if they were a race or a gender. (Toynbee, 2001)

In contrast, the social commentator, Elizabeth Kristol, in the USA, explained:

> There is a curious double standard at work here. Virtually everyone writing on medical ethics condemns abortion for the purpose of sex selection, on the grounds that it makes a powerful statement about the relative value of the female gender… But when it comes to the disabled, no such concern is apparent, since society has already sanctioned abortion for virtually any disability that the testing uncovers …
>
> The landmark [US] President's Commission on Bioethics in the 1980s illustrates this double standard. The commission gives approval for genetic testing but also condemns use of such testing for sex selection: "[Sex selection] is incompatible with the attitude of virtually unconditional acceptance that developmental psychologists have found to be essential to successful parenting. For the good of all children, society's efforts should go into promoting the acceptance of each individual – with his or her particular strengths and weaknesses – rather than reinforcing the negative attitudes that lead to rejection."[6]

It is thus necessary to determine why ableism is seen as being different to other forms of discrimination. This is important because the deselection of possible future persons who may be considered to have lives of inferior worth than others, through procedures such as heritable genome editing, may increasingly become accepted.

Selection risks and eugenics

One of the most serious concerns regarding the intentional modification of the human genome is the risk of introducing new eugenic selection, which is seen as more attractive to those encountered in the past. This is because the technology could enable the generation of children in whom specific traits are selected by parents or society, which raises concern about the future of children's commodification and a profound anthropological shift regarding the value and meaning of life.

In this context, Sufian and Garland-Thomson argue in their chapter that deselective ideology, which may be present with heritable genome editing, may be a new form of eugenic thinking grounded on what may be defined as 'common sense' ableism, which is a belief system that enables individuals to both reject any of the offensive eugenic principles of the past while actually supporting them.

Relatedly, David A. Jones in his chapter explains that selection between persons always presupposes the existence of inequalities between individuals and communities and that some of these inequalities result from biological traits. He indicates that the selection mechanism itself (whether natural selection, sex selection, selective breeding of domestic animals, or eugenic selection) does not simply presuppose inequalities between individuals or groups of living beings. Instead, it makes use of these inequalities as the very basis for selection in order to make someone or something 'better'. In other words, it requires, and thus *values,* the presence of inequalities. He also clarifies that with eugenics, 'persons' (however defined) are always unequal. Thus, eugenic procedures may not only be seen as wrong because a degree of coercion existed in some of the past reproductive practices, but also because they undermined the very basis of equality in value and worth between all existing or possible future persons in society, among whom some form of biological inequality will always exist.

Selection risks and future generations

A number of chapters in this volume, including the one from Vicente Bellver, indicate that present generations may have responsibilities towards future ones, thereby fulfilling a principle of intergenerational equality. This has already been clearly established in international agreements such as on the protection of the environment. For instance, the 1992 *Rio Declaration on Environment and Development* states: 'The right to development must be fulfilled so as to equitably meet developmental and environmental needs of present and future generations' (Principle 3). In addition, Bellver explains that international legal texts addressing obligations towards future generations often have the dual objective of protecting nature and preserving humankind.

Thus, present generations should not cause incommensurable harm to future generations. It was also noted that the exploitation of the environment was seen, in the past, as positive and useful. But with the increase in environmental pollution, this exploitation may now be considered dangerous, meaning that the environment needs protecting for future generations. Similarly, it could be argued that the human genome could be exploited and used in an inappropriate way, meaning that future generations may also need protecting from dangerous 'genomal pollution'.

Selection risks and the inclusive society

It is possible to suggest that heritable genome editing may be acceptable in well-defined circumstances. For example, Richard Playford concludes in his chapter that, providing any decisions are carefully and cautiously considered while taking into account issues of equality and justice, amongst other important principles, there is no reason for heritable genomic editing not to be a force for good in the world.

However, Calum MacKellar argues in his chapter that only when parents, themselves, would be unable to cope (for some reason) with a disabled child in an insufficiently inclusive society, would the selection of possible future children not undermine the equal value and worth of human beings. But if this was not the case, and the parents' selection was solely based on their preferences for certain kinds of children, then such a decision would undermine equality. In discussing this, MacKellar examines the expressivist argument, which states that the decision to select between possible future persons may express (in certain circumstances) an undisclosed moral position or worldview, such as a discriminatory value system, just as clearly as if this had been revealed more directly through an oral or written declaration. In addition, because of this discriminatory attitude, the decision to select may affect the lives of others in society, often in ways, and to an extent, which cannot be foreseen.[7]

This means that personal opinions about worth, meaning and equality of human life matter to the whole of society.[8] For example, Hofmann argues:

> [T]he harm of importance in the expressivist argument may not be the harm to individuals or groups with a certain characteristic or trait, but the harm to us as a society. Accordingly, the expression of importance in the expressivist argument may not only be what (vulnerable) groups perceive, but as much the social norms and values expressed by our social institutions and practices. A good society is not necessarily a society where there are no dangers, where everybody is born according to given norms of beauty or intelligence. (Hofmann, 2017, p 514)

In other words, new selective procedures, ensuring that persons with a disability do not come into existence, may increasingly challenge the whole of society to accept the equal worth and value of all persons, including those who already exist with a disability. Conversely, if all persons, including all possible future persons with or without a disability, are considered absolutely equal in value and in worth, then no reason would ever exist for any selection to take place between these future persons through heritable genome editing in a fully inclusive society.

Interestingly, the very basis of equality in civilized society is inherently based on a kind of expressivist concept of equality. Indeed, it is the belief in the inherent equal value and worth of all individuals in society that is reflected in *acts* (and not just oral and written *declarations*) of justice, fairness, and impartiality.

Finally, it is worth noting that, in a material world, there will never be a time in the future when all individuals are *biologically* equal. Moreover, what may be considered as genetic (or any other kind of) perfection will always remain a dream. Consequently, a more appropriate way forward for society may not be to fatalistically accept heritable genome editing but to strive, instead, for ever more inclusivity. In fact, a civilized society will always consider all individuals – regardless of whether they have heritable biological disabilities or differences – as equally valuable and worthy. Moreover, such a society will always seek to improve the provision of the social resources and support necessary to parents to enable them to consider having even the most different and most challenging of children.[9]

What is the right way forward?

In her chapter, Britta van Beers explains that the underlying philosophy of the UN Universal Declaration of Human Rights is that we are all born *equal in dignity, worth, and value.* Accordingly, it was in order to address the risks to equality of selective and eugenic reproductive procedures associated to unacceptable discrimination, that a number of international legal provisions opposing such selection, amongst other reasons, were prepared.[10] Laura Jacobs and van Beers mention in their chapters, that the UNESCO International Bioethics Committee had clearly indicated that, if any intentional heritable selection was accepted, this would '*jeopardize the inherent and therefore equal dignity of all human beings and renew eugenics, disguised as the fulfilment of the wish for a better, improved life*' (UNESCO International Bioethics Committee, 2015).[11] In other words, heritable genome editing could undermine the equality in worth and value of all human beings – an equality which is the very basis of civilized society.

But as the abuse which took place in the first half of the 20th century slowly becomes an ever-older memory, pressures are now returning for new forms of eugenic selection. For example, American Nobel Prize Laureate and co-discoverer of the structure of the DNA molecule, James Watson (1928–2025) wrote, in 1995, concerning the German politician and despot Adolf Hitler (1889–1945):

> But diabolical as Hitler was, and I don't want to minimize the evil he perpetuated using false genetic arguments, we should not be held in hostage to his awful past. For the genetic dice will continue to inflict cruel fates on all too many individuals and their families who do not deserve this damnation. Decency demands that someone must rescue them from genetic hells. If we don't play God, who will? (Watson, 1995, p 197)

However, it is possible to question whether a frightening dystopian future may not arise, instead, with an increased acceptance of selection procedures between possible future children, such as with heritable genome editing. This is because serious and unprecedented concerns arise when the inequality in value and worth of, and preference between, the lives of future and existing children is accepted. It would then be possible to ask whether the concept of absolute equality and the goal of full inclusivity of all persons in society, as well as the virtue of compassion, would survive. These are the questions now facing humanity.

Notes

1 Darnovsky and Hasson also explain that many of the models for public engagement that were employed or proposed are inadequate because they limit, devalue, or undermine genuine participation of diverse publics (Darnovsky and Hasson, 2024, p 57).

2 See chapter from Katie Hasson.

3 As the House of Lords Medical Ethics Select Committee in 1994 indicated, belief in the equal special worth of human life is at the heart of civilized society. It is the fundamental value on which all others are based, and it is the foundation of both law and medical practice. House of Lords Select Committee Report on Medical Ethics, 1994, paragraph 34.

4 See chapter from Katie Hasson and Trevor Stammers.

5 If the decision seeks to be rational it must be grounded on what exists in the real world. This contrasts with irrational decisions which may be based on pure chance or imaginary made-up settings.

6 This quote from the President's Commission on Bioethics was cited in Kristol, 1993, p 23, which was itself mentioned in Rae, 1996, p 209.

7 For instance, the Chief Executive of the UK Cystic Fibrosis Trust explained, in 2003, that amongst families at risk of having children with Cystic Fibrosis, data relating to pre-implantation diagnosis or pre-natal tests to avoid a child with such a disorder was difficult to obtain since: 'Families obviously feel that a child already born affected by Cystic Fibrosis may feel unwanted if they know their parents have made a decision of this nature in relation to a subsequent pregnancy' (Barnes, 2003).

8 The interests of the individual cannot always be separated from the interest of society as a whole (House of Lords Select Committee Report on Medical Ethics, 1994, paragraph 237).

9 For a development of this proposal, see MacKellar, 2021.

10 See, for example, the Parliamentary Assembly of the Council of Europe which recommended (Recommendation 2264) that the Committee of Ministers call upon the 46 member States (including the UK) to 'embrace a clear and total prohibition on establishing a pregnancy with germ-line cells, their precursors or human embryos having undergone intentional genome editing of their nuclear DNA, by introducing legislation at the national level and opposing permissive regulation at European and international levels' (Parliamentary Assembly of the Council of Europe, 2023).

11 It should be noted that a decision not to have a certain kind of child, even by deciding to no longer procreate in a natural manner, may be based on other factors, such as a genuine psychological, financial and material inability by some parents to cope with a very seriously disabled child. Such a decision would not then be eugenic in nature (and would not undermine equality) since the decision would not be based on the biological trait of the child but on the *capacity of the parents* to look after such a child.

References

Archard, D. (2004) 'Wrongful life', *Philosophy*, 79: 403–420.

Barnes, R. (Chief Executive: Cystic Fibrosis Trust), *Letter to the Scottish Council on Human Bioethics*, 22 July 2003.

Baylis, F. (2019) *Altered Inheritance*, Cambridge, MA: Harvard University Press.

Cicero (1999) *De Legibus* (On the Laws) in J. Zetzel (ed) *On the Commonwealth and On the Laws*, Cambridge: Cambridge University Press.

Darnovsky, M. and Hasson, K. (2024) 'Untangling CRISPR's twisted tales', in N. Baer (ed), *The Promise and Peril of CRISPR*, Baltimore: Maryland, Johns Hopkins University Press, pp 45–70.

German National Ethics Council (2003) 'Position in favour of the retention and more precise specification of the ban on PGD', in *Opinion: Genetic Diagnosis Before and During Pregnancy*, Berlin: Nationaler Ethikrat.

Global Observatory for Genome Editing (2025), 'Summit Statement', Available from: www.summit.global-observatory.org/statement/ [Accessed 11 June 2025].

Green, R. (1997) 'Parental autonomy and the obligation not to harm one's child genetically', *Journal of Law, Medicine and Ethics*, 25(1): 5–15.

Hofmann B. (2017) '"You are inferior!' Revisiting the expressivist argument', *Bioethics*, 31(7): 505–514.

Holm, S. (2000) 'Ethical issues in pre-implantation diagnosis', in J. Harris and S. Holm (eds), *The Future of Human Reproduction*, Oxford: Oxford University Press, pp 176–190.

House of Lords Select Committee Report on Medical Ethics, HL 21-I, 31 January 1994.

Kristol, E. (1993) 'Picture perfect: The politics of prenatal testing', *First Things*, 17–24.

MacKellar, C. (2021) 'Why human germline genome editing is incompatible with equality in an inclusive society', *The New Bioethics*, 27(1): 19–29.

Parliamentary Assembly of the Council of Europe, Heritable genome editing in human beings, Recommendation 2264 (2023), 28 November 2023, Available from: https://pace.coe.int/en/files/33182/html [Accessed 11 June 2025].

Rae, S.B. (1996) *Brave New Families: Biblical Ethics and Reproductive Technologies*, Grand Rapids, MI: Baker Books.

Reader, S. (2017) *The Ethics of Choosing Children*, Cham, Switzerland: Palgrave Macmillan.

Scott, R. (2007) *Choosing Between Possible Lives*, Portland, OR: Hart Publishing (Bloomsbury).

Song, R. (2007) 'Conclusion: Fragility and grace; theology and disability', in J. Swinton and B. Brock (eds), *Theology, Disability and the New Genetics*, London: T&T Clark, pp 234–244.

Swinton, J. (2007) 'Introduction: Re-imaging genetics and disability', in J. Swinton and B. Brock (eds), *Theology, Disability and the New Genetics*, London: T&T Clark (Continuum), pp 1–25.

Toynbee, P. (2001) 'Rights are for the living', *The Guardian*, 24 August 2001.

UNESCO International Bioethics Committee (2015) 'Report of the IBC on updating its reflection on the human genome and human rights', SHS/YES/IBC-22/15/2 REV.2, 2 October 2015, Available from: http://unesdoc.unesco.org/images/0023/002332/233258E.pdf [Accessed 11 December 2024].

Waldron, J. (2017) *One Another's Equals: The Basis of Human Equality*, Cambridge, MA: The Belknap Press of Harvard University Press.

Watson, J. (1995) 'Values from Chicago upbringing', in D.A. Chambers (ed), *Deoxyribonucleic Acid: The Double Helix – Perspective and Prospective at Forty Years*, New York: New York Academy of Science.

Wolbring, G. (2012) 'Expanding ableism: Taking down the Ghettoization of impact of disability studies scholars', *Societies*, 2(3): 75–83.

Index

T

U

V

W